DEPARTMENT OF THE ENVIRONMENT
PLANNING RESEARCH PROGRAMME

GOOD PRACTICE ON THE EVALUATION OF ENVIRONMENTAL INFORMATION FOR PLANNING PROJECTS

RESEARCH REPORT

1994

Land Use Consultants
assisted by
The Environmental Appraisal Group,
University of East Anglia

London : HMSO

First published 1994
ISBN 0 11 752990 7

Produced from Camera Ready Copy supplied by the
Department of the Environment

GOOD PRACTICE GUIDE ON THE EVALUATION OF ENVIRONMENTAL INFORMATION FOR PLANNING PROJECTS

RESEARCH REPORT

May 1994

Land Use Consultants
assisted by
The Environmental Appraisal Group,
University of East Anglia

Trioka House,	43 Chalton Street	Gleniffer House	**21 Great George Street**
2 East Union Street	London NW1 1JB	21 Woodside Terrace	**Bristol**
Rugby, Warwickshire		Glasgow	**BS1 5QT**
CV22 6AJ		G3 7XH	
Twl: 0788 553993	071 383 5784	041 332 8098	**0272 291997**
Fax: 0788 550257	071 383 4798	041 333 0918	**0272 291998**

PREFACE

The research reported in this document was commissioned by the Department of the Environment and undertaken by Land Use Consultants, assisted by the Environmental Appraisal Group at the University of East Anglia. It was overseen by a Steering Group comprising representatives of the Department of the Environment, the Scottish Office Environment Department, the Welsh Office and the Planning Inspectorate. The research was undertaken during the period January to July 1993.

The views expressed in this report, as well as the conclusions, are those of the authors and do not necessarily represent the views of the Steering Group or the Department of the Environment, the Scottish Office Environment Department, the Welsh Office or the Planning Inspectorate.

CONTENTS

APPENDICES

1.0 INTRODUCTION

1.1 The research reported in this document was commissioned to enable the Department of the Environment, the Scottish Office Environment Department and the Welsh Office to produce a good practice guide to assist local authority decision-makers in England, Scotland and Wales in their evaluation of "environmental information". This is information which must be taken into account when determining planning applications which are the subject of an environmental statement.

BACKGROUND

1.2 Directive 85/337/EEC has been implemented for projects that require applications for planning permission in England and Wales by the Town and Country Planning (Assessment of Environmental Effects) Regulations 1988[1] . Where environmental assessment (EA) is required, developers must prepare and submit in conjunction with the application for planning permission an environmental statement (ES) setting out their own assessment of the likely environmental effects of the proposed development and the measures envisaged to avoid, reduce or remedy any significant adverse effects. The ES must contain the "specified information" required by paragraph 2 of Schedule 3 to the Regulations and may include the "further information" listed at paragraph 3 of Schedule 3 to the Regulations. There must be a non-technical summary of all the information.

1.3 The ES must be sent to statutory consultees and made available for inspection and purchase by the public. The "environmental information", ie the ES and any representations on it from statutory consultees and other persons, must be taken into consideration by the decision-taking authority (usually the planning authority but may be the Secretary of State or an inspector/reporter) before planning permission may be granted.

1.4 The comparatively limited number and uneven distribution of planning projects that require EA means that first, many developers will have to prepare an ES infrequently and, secondly, most planning authorities will have to deal with only one or two EA cases a year at most, so their practical experience will be limited. In the White Paper on the Environment published in September 1990 the Government promised that it would provide guidance on how best to prepare and evaluate environmental statements. The Department of the Environment is in the process of publishing good practice guidance on the preparation of environmental statements for planning projects.

PURPOSE OF RESEARCH

1.5 The purpose of the research was to review existing literature and planning authority experience, including the use of monetary and non-monetary evaluation techniques, and prepare draft guidance to assist local authority planning officers, planning committees and others to:

(i) consider whether submitted ESs are adequate or whether additional information needs to be sought from the applicants; and

(ii) evaluate the information in the ES and any representations thereon from statutory consultees and others, so that this can contribute to an informed decision on the application for planning permission;

[1] In Scotland, the Environmental Assessment (Scotland) Regulations 1988.

1.6 The research report summarises the findings of the review of literature and experience and considers the likely nature and extent of guidance. The draft guidance is the subject of a separate document.

STRUCTURE OF THE RESEARCH REPORT

1.7 The research report comprises six sections, Section 1 of which is the introduction. Section 2 describes the method of research. The next three sections summarise the main findings under the following headings:

Section 3: The procedures used by planning authorities to handle environmental statements and evaluate environmental information.

Section 4: The quality and adequacy of environmental statements.

Section 5: Methods and techniques of evaluation and their potential use.

Section 6 summarises the main conclusions of the research and suggests areas where guidance may be helpful.

2.0 METHOD OF RESEARCH

2.1 The research was carried out in three separate but overlapping stages :

Stage 1: Review of literature and current practice.

Stage 2: Study of evaluation methods and techniques, including selected case studies.

Stage 3: Production of the research report.

STAGE 1

REVIEW OF LITERATURE

2.2 There is a growing body of literature about environmental assessment and related subject areas such as environmental economics. The review of this literature concentrated on references which were considered to be of particular relevance to the evaluation of environmental information in EA. The references were drawn from a wide range of sources including databases on EA evaluation, notably those held by the EIA Centre at the University of Manchester and by the University of East Anglia, and international sources such as the World Bank's Environmental Assessment Source Book.

2.3 The references were grouped under three headings :

(i) **General EA** : those which provide general background about the evaluation of environmental information in EA, including standard texts on EA and related subjects, as well as the handbooks prepared by Cheshire, Essex and Kent County Councils.

(ii) **Adequacy of Environmental Statements** : those which explain the techniques which have been developed to assess the adequacy of environmental statements, including the Lee-Colley Review Package, the Review Criteria used by the Institute of Environmental Assessment (IEA); also handbooks prepared by various statutory and non-statutory consultees which deal with the adequacy of ESs as well as their preparation.

(iii) **Evaluation Methods and Techniques** : those which explain the various methods and techniques for evaluating environmental information, including the application of techniques which rely on calculating a monetary value for an environmental variable.

2.4 A full bibliography is attached as **Appendix 1.** This comprises three sections corresponding to paragraph 2.3, i.e.: (1) general references; (2) references for reviewing the adequacy of environmental statements; and (3) texts relating to evaluation methods and techniques. Section 3 of the bibliography also contains short abstracts of the key evaluation texts identified during the research. The aim is to help the reader identify those texts which may be especially relevant when faced with a particular set of circumstances.

REVIEW OF CURRENT PRACTICE

2.5 The review of current practice comprised three areas of work :

- interviews with practitioners;
- telephone survey of selected planning authorities;
- selection of case studies.

2.6 **Interviews with Practitioners:** A number of practitioners in the general field of EA were interviewed. The purpose of these discussions was to consider existing practice and highlight areas upon which the research should concentrate. The following people contributed :

Local Authorities:
Dr Elizabeth Street, Principal Planning Officer, Kent County Council
Roy Leavitt, Senior Principal Planner, and Peter Hakes, Area Planning Officer, Essex County Council

Consultees:
David Coleman, South East Regional Officer, Countryside Commission
David Pritchard, Planning Unit Manager, RSPB
William Sheate, Assistant Secretary, and Petra Biberbach, Northern Policy Officer, CPRE

Universities etc.:
Professor Malcolm Grant, Dept of Land Economy, University of Cambridge
Dr Peter Hopkinson, Department of Environmental Science, University of Bradford
Professor Christopher Woods and Norman Lee, Co-Directors, EIA Centre, University of Manchester

Consultants:
Elizabeth Dower-Jeffrey, Managing Director, Oakwood Environmental Consultants
Nicola Beaumont, Senior Consultant, Environmental Resources Management
Paul Tomlinson and Brendon Barrett, Ove Arup & Partners
Mark Lintell, Chairman and Director, Land Use Consultants

Other:
James Winpenny, Research Fellow, Overseas Development Institute

2.7 **Telephone Survey of Selected Planning Authorities:** A total of 54 planning authorities were contacted by telephone to discuss how they currently deal with the question of adequacy of environmental information and the process of evaluation. The discussions were based on a standard list of questions to ensure a consistent and structured approach.

2.8 The aims of the telephone conversations were to :

- establish how planning authorities decide whether ESs are adequate or whether additional information is required;

- determine how planning authorities evaluate the information in an ES to reach a decision;

- identify possible projects for detailed investigation as case studies.

The summary of the findings from the telephone survey is included as **Appendix 2**.

2.9 **Selection of Case Studies:** Projects appropriate for detailed analysis as case studies were identified during the telephone interviews with planning authorities.

2.10 The case studies were selected to reflect one or more of the following criteria:

- an even geographical spread throughout England, Scotland and Wales;
- different types of planning authority (i.e. District Council, Metropolitan Borough Council, London Borough Council, County Council, Regional Council and National Park Authority);
- different types of planning project subject to EA;
- planning applications which have been decided (i.e. approved or refused; and
- rural and urban locations.

2.11 It was initially proposed that 20 case studies would be undertaken. It became apparent upon the completion of the first six, however, that they were not providing as much useful material on the evaluation of environmental information as had been anticipated. Consequently, it was decided not to complete all 20, but rather to complete 11 and then to explore one example in more detail to ascertain how monetary evaluation techniques might have been employed when analysing environmental information. The completed case studies are listed in Table 2.1 and the detailed review of one of these, a proposed city airport, is written up in **Appendix 5**.

STAGE 2

STUDY OF EVALUATION METHODS AND TECHNIQUES, INCLUDING SELECTED CASE STUDIES

2.12 For the purpose of this research report a distinction is made between '**techniques**' which are used in specialist fields to evaluate the importance of individual environmental, social and economic topics, and '**evaluation and decision-making methods'** which seek to aggregate information from all sources and draw conclusions about their relative importance. Many of the techniques for evaluating information which have been examined as part of this research are described in the literature as decision-making methods, although they are in practice very often capable of making only a partial contribution to the overall decision-making process. This is the case with most of the monetary and non-monetary techniques described in **Appendices 3 and 4.**

2.13 The review of evaluation methods and techniques concentrated on those which identify and weigh, in some form, all the relevant issues, including environmental issues, which are important for the decision on a planning application. There is a considerable and growing volume of literature on the methods and techniques which exist, although only a few references include examples of how they have been used in planning practice.

2.14 One of the purposes of the case studies was to probe the issue of practical application further and ascertain whether any of the methods and techniques available had been used by planning authorities to assist them in reaching a decision on a planning project supported by an ES. The case studies also provided the opportunity to :

* establish, through contact with the planning authorities, how and at what stage the local planning authority and developer determined what environmental information should be included in an ES;

* see how the planning authority officers set about reviewing the information in the ES and assessing its adequacy; and

* see how the "environmental information" was presented by planning officers to decision-makers.

TABLE 2.1 - CASE STUDIES

TYPE OF PLANNING AUTHORITY	CASE STUDY
District (England)	Mixed use development comprising hotel, conference centre, golf course, equestrian facility and timeshare properties.
District (England)	Major sport and leisure complex sponsored by the local authority.
District (England)	Car factory
District (Scotland)	Extension to a quarry.
District (Wales)	Pharmaceutical manufacturing plant including waste incinerator.
Metropolitan Borough	City airport.
London Borough	Redevelopment of a small arms factory for mixed use.
County (England)	Extension to a landfill site.
County (England)	Route improvement to a single carriageway road.
County (England)	Extraction and processing of sand and gravel.
National Park	Extension to a limestone quarry.

2.15 A summary of the key points to emerge from the case studies is included in **Appendix 6**.

STAGE 3

PRODUCTION OF THE RESEARCH REPORT

2.16 The research report pulls the various threads of the research together and identifies areas where guidance may be useful. A great deal of information was collected, most of which is summarised in the Appendices. However, much of the detailed information obtained from the telephone survey of local authorities, the meetings with practitioners and the case studies is not reproduced. This is for reasons of confidentiality and brevity.

171733 B

3.0 THE PROCEDURES USED BY PLANNING AUTHORITIES TO HANDLE ENVIRONMENTAL STATEMENTS AND EVALUATE ENVIRONMENTAL INFORMATION

3.1 Most research to date on the operation of the EA planning procedures has focused on the adequacy of ESs submitted under the Regulations. Relatively little attention has been paid to the capacity of planning authorities (in terms of both time and experience) to handle ESs, or their actual performance in vetting and evaluating them.

3.2 From the telephone interviews, the selected case studies and the comments received from a number of practitioners, it is apparent that the various approaches adopted by planning authorities for dealing with environmental statements have six common elements :

(i) **Pre-application discussions/scoping:** The purpose of these discussions is to identify the key issues and agree the nature and scope of an ES before too much baseline survey and analytical work has been carried out by the applicant. Pre-application discussions and scoping are, however, only possible where a planning authority knows about the proposal at an early stage in its development. Nineteen authorities (35%) interviewed on the telephone actively seek to do this, and a third of these provide a detailed checklist of the topics to be covered by the applicant. A further five authorities (9%) complete an in-house scoping exercise either before or after the planning application is submitted, independent of discussions with applicants.

(ii) **Review of the Content and Quality of an ES:** Although at first sight there appear to be two distinct elements to reviewing the content and quality of an ES (ie. determining the adequacy of the information contained within an ES, and evaluating the information to come to a decision), in practice the two are often indistinguishable. An initial review of the content and quality of an ES is undertaken by most local authority planning officers once an application is submitted. Such reviews may either take the form of a simple checklist or can extend to the use of more complex techniques such as the Lee-Colley Review Package[1] . Evidence from the case studies suggests a simple checklist is the most commonly used approach. However, two of the case study authorities have made use of the review service provided by the Institute of Environmental Assessment, one of them being on a regular basis.

One third of the authorities interviewed on the telephone have used consultants with acknowledged expertise in EA to evaluate ESs on occasions. Some authorities which had not used consultants said that they might do so in the future, although there was a general feeling of concern that resource constraints may prevent this happening in practice. Some authorities stated that they could not afford consultants in any circumstances.

In addition to using Schedule 3 to the EA Regulations as a checklist, published guidance such as the Department of the Environment's 'Blue Book' (Environmental Assessment : A Guide To The Procedures, 1989) and Circulars[2] was identified as being of some help in reviewing the content and quality of ESs.

---

[1] Lee, N. & Colley, R. (1990). Reviewing the Quality of Environmental Statements - EIA Centre, University of Manchester - Occasional Paper No.24.

[2] DOE Circular 15/88 (Welsh Office Circular 23/88) Environmental Assessment.
SDD Circular 13/88 Environmental Assessment: Implementation of EC Directive: The Environmental Assessment (Scotland) Regulations 1988

(iii) **Consultation with Statutory and Non-statutory Consultees:** The EA Regulations list those public bodies with statutory environmental responsibilities who must be consulted on the submitted ES. Obtaining information and comment from experts in particular subject areas is an important part of the evaluation process, not least because it provides additional environmental information. It was evident from the telephone survey that planning authorities tend to rely quite heavily on statutory consultees and, to a lesser extent, other consultees, to help assess and evaluate the information provided in an ES. Thirty-three (61%) of the authorities contacted on the telephone stated that it is an important part of the process, although some planning authorities stressed that this would usually be the case also for applications not requiring an ES.

The reliance placed by planning authorities on statutory consultees for their opinion on the ES gives cause for some concern. For example, certain planning authorities complained that statutory consultees are too parochial in their assessment of ESs, only looking at the areas within their remit and not paying enough attention to the wider issues raised by the application. This is not always helped by the practice of some planning authorities of only sending for comment those parts of the ES relating to the area of expertise of the statutory consultee in question. Other planning authorities felt that some of the statutory consultees failed to provide sufficient guidance on what are often highly technical areas which, although often governed by other environmental protection legislation, may have a considerable bearing on the final planning decision.

The research has also shown that in some instances contact between the planning authority and statutory consultees has simply involved transmitting a copy of the ES and asking for comments, as occurs with any reasonably sized planning application. This has tended to elicit general responses rather than comments focused on key issues and specific areas of concern to the statutory consultee.

Many authorities found the advice provided by non-statutory consultees useful. The organisations most often mentioned were the county wildlife trusts, although national bodies such as the RSPB have also given helpful advice on occasions.

(iv) **Public Consultation:** The views of the general public may also be an important sourcc of environmental information, as well as assisting in the evaluation process. The extent of public consultation appears to bear a strong relationship to the sensitivity of an application, the more sensitive ones provoking greater public interest (and controversy) and resulting in a desire amongst local authorities for consultation over and above the statutory requirements. Eight authorities interviewed on the telephone (15%) claimed that they encouraged an applicant to hold public meetings or an exhibition.

(v) **Evaluation of the Environmental Information:** The evaluation of the environmental information to assist in reaching a recommendation on a planning application tends to occur once all the information has been received. This issue was discussed in detail with the case study officers, the majority of whom stated that they appraised a planning application accompanied by an ES in the same way as any other application. One or two of the local authority planning officers interviewed went as far as to say that the ESs did not contribute greatly to the decision-making process. In their view, this poor contribution partly reflected the inadequacy of the ESs but was mainly because the negotiations which took place after the planning application was submitted had more influence on the decision than the content of the ES itself.

(vi) **Presentation of Information to Decision-makers:** The presentation of all relevant information to the decision-maker(s), usually the Planning Committee, is the final stage for the planning authority in dealing with an application (assuming it does not go to appeal). This issue was also discussed in detail with the case study officers. For each scheme examined, the report to Committee comprised the following elements in some form :

- description of the proposal;
- planning history/background;
- planning policy context;
- summary of consultations, representations and objections;
- officer's discussion/appraisal;
- officer's recommendation and suggested conditions.

3.3 Most of the authorities which have dealt with a relatively large number of ESs (in some cases 12 or more) have developed very effective procedures for steering an ES through the development control process, reflecting in some form the six common elements described above. However, it is also apparent from the research that some authorities are ill-prepared to deal with an ES when it first arrives. Junior officers who are often the first to examine the planning application and ES, are sometimes unaware of the requirements of the Regulations, and unsure how to proceed with examining and processing the ES, let alone evaluate its contents.

3.4 During the telephone interviews, fifteen (28%) authorities said that at least one member of staff in the planning department had been on training courses covering EA, although there was some concern that general courses do not provide the detailed coverage required. Ten (19%) authorities expressed particular concern about the lack of time available to assess an ES and a similar number believed that they lacked the expertise to deal with ESs.

3.5 The research suggests that a common reaction amongst development control staff to a lengthy and apparently competent ES is to assume that the critical environmental issues have been addressed. There appears sometimes a nervousness to appraise critically the contents on the grounds that it is a specialised area of work in which the officers have little knowledge or experience. Comments from some local authority planning officers suggest that some inadequate ESs have been accepted, processed and used as the basis for decisions.

3.6 In summary, it appears that those planning authorities which have had to process several planning applications accompanied by an ES are already doing so satisfactorily, whereas less experienced authorities are not so well organised. This finding from the research highlights that the guidance should be targetted at the less experienced authorities and should concentrate on how existing procedures might be improved.

4.0 THE QUALITY AND ADEQUACY OF ENVIRONMENTAL STATEMENTS

4.1 Some divergence of opinion exists amongst practitioners from both planning authorities and consultancies about the degree of attention which should be paid to the quality of ESs, as opposed to the adequacy of the overall information base on which the planning decision is eventually founded. The literature and experience relevant to assessing the quality and adequacy of ESs illustrates this dichotomy. Seven main sources were looked at, of which the first six represent the 'quality' approach and the seventh, the 'overall adequacy' approach

(i) **The Lee-Colley Review Package :** The Lee-Colley Review Package (named after the two principal authors) was published as Occasional Paper No. 24, by the EIA Centre at the University of Manchester in 1990. This package is designed to be usable by reviewers who may not possess specialist environmental expertise but who are familiar with the relevant EA regulations, and have at least a basic non-specialist knowledge of EA methodologies and current ideas on good practice in EA.

The EIA Centre decided that a self-contained review package would be most useful and that this should contain advice for reviewers, a list of the criteria to be used in each ES review and a collation sheet on which to record the findings from using the criteria. A grading system ranging from ' grade A - relevant tasks well performed' to 'grade F - very unsatisfactory' is used to assess the degree to which each criterion is met. These individual grades are then amalgamated to produce a final grading, with ESs graded at 'C' or above considered to be broadly satisfactory. Although this tends to be the method most commonly associated with reviewing the quality and adequacy of an ES, only a handful of the planning authorities interviewed on the telephone have attempted to apply Lee-Colley in practice. **Appendix 7** provides further information on the Lee-Colley methodology.

(ii) **Review Criteria**[1] **:** Another approach is to use a standard checklist, or 'Review Criteria', as advocated by Tomlinson (1989). The thinking underpinning the 'Review Criteria' approach bears comparison with that for Lee-Colley. As argued by Tomlinson (1989) :

"... judgements regarding the adequacy of an ES should be based upon a set of consistent and well defined criteria. Such criteria ought to be logically organised, allowing judgements of performance of individual components of an ES, such as baseline description, impact identification, prediction, evaluation, mitigation and monitoring."

With this in mind, Tomlinson has put forward criteria to aid the review process, and these are reproduced in **Appendix 7**.

(iii) **The Institute of Environmental Assessment (IEA) Review Criteria:** Published in 1991, the Institute of Environmental Assessment (IEA) Review Criteria are largely based on the Lee-Colley Method. Planning authorities belonging to the IEA are entitled to a review by the IEA of one ES free of charge each year, with further reviews charged on a sliding scale. Only one of the case study authorities has regularly made use of the review service provided by the IEA.

---

[1] Tomlinson, P. (1989). Environmental Statements: Guidance for Review and Audit. The Planner, 3rd November 1989.

(iv) **The Environmental Assessment Handbook of English Nature, the Countryside Council for Wales and Scottish Natural Heritage:** Some statutory consultees for EA have also adopted the Lee-Colley method as the preferred approach. English Nature, the Countryside Council for Wales and Scottish Natural Heritage use this procedure, albeit in a slightly modified form, in their joint Environmental Assessment Handbook (1992). The same grading system is used to review ESs and the full Lee-Colley Review Package is appended to the handbook. The handbook argues that best use of resources would indicate a two-stage review, given there is often little time to respond to an ES, and the only information available to the reviewer may be that presented in the ES itself. The first stage should be a non-technical review to ensure that the ES presents the necessary range of information; the second stage should be a technical review to examine the results of any specialist studies and the conclusions which are based on them.

(v) **The Countryside Commission's guidance note 'Environmental Assessment: The treatment of landscape and countryside recreation issues':** When reviewing ESs, the Countryside Commission recommend reference to Tomlinson's Review Criteria and particularly to the Lee-Colley Review Package. Their EA guidance note includes a checklist of review criteria which are grouped in the same way as Lee-Colley.

(vi) **The CPRE's pamphlet 'Environmental Statements - Getting Them Right':** Believing the quality of ESs in the UK to be generally poor, CPRE have produced guidance on producing and reviewing such documents. Their pamphlet "Environmental Statements : Getting Them Right" (1990) is intended to be a guide to the principles of good ESs and to act as a checklist for planners, developers and environmentalists alike. The pamphlet focuses on those issues that CPRE believe are too frequently dealt with poorly or inadequately. These fall into the three main areas of identifying environmental impacts, assessing such impacts and involving the public in the EA process.

(vii) **The approach of Kent County Council**: An alternative approach has been advocated by Street (1993) and others. Based on the premise that 'criteria-based' approaches essentially involve a snapshot view of an ES at a particular point in time (usually upon initial receipt of the ES), the approach seeks continually to expand and amend the submitted ES over time as more and more information is made available. The emphasis is placed on the overall assemblage of information which is presented to decision makers rather than the content and quality of the submitted ES. The approach reflects the fact that planning authorities may not refuse to process a planning application on the grounds that the quality of an ES is unacceptable; if the ES provides the "specified information", it has to be processed like any other planning application. The eventual reporting of an environmental assessment to a planning committee is a synthesis of consultees' views, representations, professional judgements and the results of any additional work undertaken whilst processing the planning application, rather than merely a presentation of the ES. Consequently it is more appropriate to assess quality and adequacy once the planning application has been processed rather than at the outset.

4.2 Most of the literature on methods for reviewing ESs has concentrated to date on the quality of the written statement without necessarily exposing the adequacy of the information contained within. When such an emphasis on the quality of the ES occurs in practice, this can lead to objectors and members of planning committees being highly critical of the Statement and ignoring the fact that much supplementary information may have been provided in other documents including the planning application itself, or may have been collected subsequently to the ES being submitted.

4.3 Two schools of thought exist about the standards which should be required from the initial ES. The first view maintains that the EC Directive and the EA Regulations require developers to undertake an environmental impact assessment which provides, as a minimum requirement, certain basic levels of information. Developers should therefore be expected to produce ESs of a high standard from the outset. In line with this belief, obvious shortcomings or misleading information should be used by planning authorities as grounds for rejecting the document until such shortcomings have been addressed. In support of this view, one or two of the case study authorities cited examples where they considered the existing procedures had placed an unreasonable burden on staff in requiring them to carry out extensive reviews of inadequate material, and to press for supplementary information. Further, some planning authorities were of the opinion that where an ES is judged to be inadequate, they should be entitled to refuse to register the planning application.

4.4 Another view is that, by the time a full appraisal of the submitted information has been carried out and the views of the statutory consultees and the planning authority's own assessment have been taken into consideration, the original ES may no longer be the most relevant document and its quality should not really affect the decision. Supporters of this view therefore argue that the present procedures could be greatly improved by requiring a revised statement to be produced on completion of the planning authority's appraisal, either by the developer, or by the authority itself. If by the authority, this document might consist of the factual sections of the planning officer's report to committee, summarising all the information and views gathered during the course of the evaluation. One problem with this, however, is that the suggestion of a 'revised statement' is likely to be seen as increasingly regulatory, particularly if the developer was expected to produce it.

4.5 Both viewpoints have some substance although, as indicated, the suggestion of a 'revised statement' on completion of the planning authority's appraisal is likely to be seen as over regulatory. We would advocate that the objective of the whole EA system should be the production of ESs of a high standard at the outset, whilst retaining the option to reject a document if it displays obvious shortcomings or contains misleading information. Where there are deficiencies in an ES, it is important that these are identified and rectified as soon as possible. The Regulations empower a planning authority to require the developer to provide specified further information, and the authority shall not determine the application until the further information is provided. It may be that these provisions should be extended to encompass the quality and adequacy of the analysis within the ES as well as the information provided.

5.0 METHODS AND TECHNIQUES OF EVALUATION AND THEIR POTENTIAL USE

THE EVOLUTION OF METHODS AND TECHNIQUES OF EVALUATION

5.1 There is considerable interest and a significant amount of research effort going into both monetary and non-monetary techniques for valuing the environment by government departments, public agencies, private companies and research institutions. The DOE publication 'Policy Appraisal and the Environment' (DOE 1992) describes how to assess environmental costs and benefits in the appraisal of policy which impacts upon the environment.

5.2 More specifically, the report of the Standing Advisory Committee on Trunk Road Assessment (SACTRA) on 'Assessing the environmental impact of road schemes' (Department of Transport 1992a) describes the development of the evaluation methods used for road schemes, discusses the adequacy of these methods and gives recommendations for their improvement. Another report for the Department of Transport by the Rendel Planning and Environmental Appraisal Group (1992) considers both monetary and non-monetary evaluation techniques in the context of the environmental appraisal of trunk road schemes in the UK. In addition, a research project has recently been conducted to assess whether the various techniques available to value non-market environmental effects are applicable in the context of Scottish Office and Scottish Enterprise interests (Hanley, 1990).

5.3 The research which is the subject of this report has revealed that there are some significant areas of uncertainty and misunderstanding amongst local authority planning officers about the EA process and what is meant by evaluation of "environmental information". The uncertainty and apparent misunderstanding appear to have resulted in the resistance of planning officers to the widespread use of methods and techniques. Nevertheless, the review of the various methods and techniques that are available has served to illustrate certain elements which could usefully be applied in practice. These divide into two broad areas :

(i) Assessing the quality and adequacy of an ES.

(ii) Evaluating environmental information as part of the decision-making process.

5.4 With regard to (i) above, methods and techniques available to assist in assessing the quality of an environmental statement are discussed above in Section 4. With regard to (ii), the following paragraphs consider the methods and techniques available to assist in evaluating environmental information as part of the decision-making process.

5.5 Environmental assessment and evaluation may call for the assembly of a wide range of information, covering many different fields and specialisms. This information needs to be sorted and prioritised, and its relative value determined as a basis for reaching a decision on whether or not the proposed development should proceed. The questions which lie at the heart of this research are concerned with the ways in which such information should be organised and presented; in other words, what methods or techniques can be used to identify the key issues and assist the decision-maker in reaching a balanced conclusion.

5.6 Decisions about development projects need to have regard to a range of environmental, technical, economic and social factors. A number of techniques have been developed for evaluating the different kinds of information relevant to these, as illustrated in the following examples:

(i) Natural habitats may be classified in terms of their nature conservation value, using criteria like extent, species diversity, and rarity;

(ii) Technical efficiencies may be measured in terms of energy consumption;

(iii) Economic returns may be assessed as a balance of costs and benefits;

(iv) Social effects may be judged using attitudinal surveys and census-type information.

5.7 While such techniques may be useful in determining the relative merits of a number of options and determining preferences within a single discipline they do not provide a convenient way of aggregating information of different types, or of weighing the relative significance of this different information in order to arrive at an overview. To achieve this goal some broader framework for decision-making is required.

5.8 As stated in paragraph 2.12, for the purpose of this report a distinction is made between 'techniques' which are used in specialist fields to evaluate the importance of individual environmental, social, and economic topics, and 'evaluation and decision-making methods' which seek to aggregate information from all sources and draw conclusions about their relative importance.

5.9 Examination of the literature reveals that the impetus to develop some convenient framework for decision-making on major projects has come from three main areas of practical and academic research. These are engineering; economics; and planning, design and environmental sciences; with business and project management making an input in each of these fields. Inevitably there has been considerable overlap and cross-fertilisation between disciplines in the search to improve methods and techniques and to produce a universally acceptable approach to decision-making. The contributions from the three principal disciplines are examined below. Passing reference is made to some of the evaluation and decision-making methods which have evolved. These are described more fully in paragraph 5.18 and **Appendix 3.**

5.10 **Engineering:** Major engineering development programmes have generated a demand for procedures which can simplify areas of choice between alternatives and options. The US Corps of Engineers was particularly influential after the second world war in developing techniques like Pair-Wise Comparison, for selecting between alternative dam sites in the Tennessee Valley, and for use in flood-routing schemes. Such techniques were combined with systems analysis for modelling major river catchments, and in the 1960s and 1970s great faith was placed in the power of computer modelling to generate the most efficient engineering solutions to complex resource development projects. The omission of the human dimension in such methods prompted widespread criticism from environmental groups and stimulated the demand for environmental impact assessment in the United States, paralleled by the creation of the National Environmental Protection Agency in 1969.

5.11 **Economics:** A close link has always existed between engineering and economic inputs to major development projects, especially in the case of those in the public sector where there has been a need to justify government funding, such as the example already quoted of reservoir development in the United States. Cost-Benefit Analysis (CBA) was developed as a comprehensive method for measuring the direct costs and benefits of promoting such schemes and it has remained at the forefront of economic analysis for over 60 years. However, CBA came under increasingly heavy criticism during the 1970s as growing environmental awareness revealed that public concerns about the polluting impacts of major development were not being adequately covered in the costs, while intangible environmental qualities were being excluded from the benefits. Recent appraisals of the usefulness of CBA have been undertaken by Jacobs (1991) and Dixon and Sherman (1991).

5.12 In spite of these criticisms CBA has remained one of the strongest economic tools and progress in the development of monetary valuation techniques has made it increasingly able to take account of environmental cost and benefits. Where monetary valuation techniques cannot be applied, the usual procedure is to set out physical data on environmental impacts alongside the information on economic costs and benefits. The results of EA can be used in this manner.

5.13 In the light of the criticisms made of CBA, efforts were made in the 1970s to develop a method which would incorporate non-monetary values alongside financial costs when considering alternative development plans. The method known as the Planning Balance Sheet, was produced by Nathaniel Lichfield, an economist and planner, and represented one of the first attempts to combine economic theory, with the practical requirements of town planning.

5.14 **Planning, Design and Environmental Sciences:** The disciplines of planning and landscape design, supported by the general principles of ecology, have played an important role in the development of planning methods for evaluating environmental options. McHarg[1], an American landscape architect, is credited with developing one of the first comprehensive methods for using overlays to select preferred development sites. This method was used extensively in early EAs, for example in assessing pithead sites for the Vale of Belvoir Coalfield in the 1970s, and is still used as a site selection technique, although it is no longer credited with having the capacity to incorporate all aspects of the decision-making process.

5.15 The ideas initiated by McHarg were picked up by environmental scientists who started to apply systems and network theory to environmental assessment, and to develop evaluation matrices. The 'Leopold Matrix'[2] was used in many early US environmental impact assessments, and attempts were made to introduce ranking, scoring and weighting in order to assess the relative importance of different types of environmental impact, for example on landscape and ecology. Certain of these approaches, including the Batelle Laboratories Method[3] , attempted to reduce all environmental effects to a single index figure. The use of mathematical models of this type has been generally discredited, although Sonntag[4] and others continue to use derivatives of this method for evaluating specific topics, The principal criticism of weighted matrices, checklists and networks is that they reduce complex areas of environmental concern to single figures or formulae which conceal from the decision-maker the various assumptions which have been made and the trade-offs between different types of impact.

---

[1] McHarg, I.L (1969) Design with Nature. The Natural History Press

[2] Leopold L.B., Clark F.E., Hanshaw B.B., & Balshey J.R. (1971). A Procedure for Evaluating Environmental Impact. US Geological Survey Circular 645, Dept. of Interior, Washington D.C.

[3] The Environmental Evaluation System (EES) was developed by the Battelle Laboratories of Columbus, for the US Bureau of Land Reclamation in the early 1970s. It was specifically intended for use in the assessment of water resource projects, but is potentially applicable to other types of development.

[4] Sonntag, N and others (1980). Integration: a role for adaptive environmental assessment and management. In Proceedings of the Symposium on Effects of Air Pollutants on Mediterranean and Temperate Forest Ecosystems, Report PWS-43.

5.16 Most, if not all, of the methods and techniques outlined above have been shown to have value in particular circumstances but the search continues for a single comprehensive framework which can draw together the best elements of CBA, pair-wise comparison, overlays, EA matrices and checklists.

5.17 There has also been some interest in applying operational research and decision analysis to the difficult value judgements which are involved in EA. This is a logical development in that the issues which have to be resolved are no different from other areas of decision-making involving the weighing of competing values.

5.18 An examination of the three main areas of practical and academic research that have provided the impetus to develop a convenient framework for decision-making for major projects, highlights a number of methods which have received some recognition in practice. The most commonly used are summarised below (and reviewed in more detail in **Appendix 3**), followed by an explanation in paragraph 5.19 of the principal monetary evaluation techniques upon which some of them in part depend if they are to be useful.

(i) **Cost benefit analysis:** Cost-benefit analysis (CBA) is based on the principle that an investment project should only be undertaken if all its benefits outweigh all its costs. If a project has several alternative forms, or there are competing projects, the one with the highest benefit cost ratio should normally be chosen. As costs and benefits (including those of an environmental nature) can only be added and subtracted if expressed in the same units, CBA attempts to place monetary values on them, money being a convenient 'measuring rod of value'. Whilst it can be a useful way of weighing environmental considerations in the balance with economic considerations, there are problems in the use of monetary valuation techniques. In particular, they can be time consuming and expensive in their use of skilled personnel; and they can result in estimates which convey an unrealistic level of accuracy.

(ii) **Planning Balance Sheets:** In the light of the criticisms made of CBA, a method known as the Planning Balance Sheet (PBS) was developed by Nathaniel Lichfield. Variables that are capable of being assessed in monetary terms are entered into the Balance Sheet, as are costs and benefits that defy monetary quantification or are quantifiable in monetary terms but due to lack of time, expense, or lack of data are not measured in this way. Where an improvement over the existing situation is envisaged for those variables not quantified, a positive sign (+) appears in the PBS and, conversely if a deterioration is predicted, a negative (-) sign is used. Whilst this approach avoids the use of monetary valuation techniques, it does not fully address the extent (as distinct from the direction) of environmental impacts.

(iii) **Goals achievement matrix:** The Goals-Achievement Matrix (GAM) was developed by Morris Hill after criticisms that both CBA and the PBS tend to classify costs and benefits with no regard to the politics of a particular situation. The constituent elements of a plan or project, such as housing, employment or open space are weighted before the matrix is constructed to indicate the political preference or priority attached to these variables. Whilst the objectives and value judgements of the decision makers are explicitly stated with this method, determining whether or not a plan or project moves towards or away from a defined goal (which may conflict with other goals or be ultimately unattainable) could have considerable resource implications.

(iv) **The Integration of Environmental Assessment into Planning:** This is a recent proposal from Nathaniel Lichfield and represents a modified form of the

PBS. The method requires both the input of the physical environmental assessment data, and the results of social surveys which seek to measure the attitudes and responses of different sectors of the local community to a development proposal, and either the 'do-nothing' situation or any alternative course of action. This approach requires considerable resources to operate and also suffers from some of the difficulties associated with non-monetary evaluation techniques when it comes to identifying the relevant groups in the community.

(v) **Pair-wise comparison:** Pair-wise comparison was first developed by the US Army Corps of Engineers and is a procedure that attempts to progressively rank the desirability of alternative options, by directly comparing them against each other. The criteria considered to be of importance when evaluating and comparing the different options may be limited to environmental factors or may include technical and economic factors. Pair-wise is most suited to comparing different options, such as alternative routes for a bypass or alternative sites for waste disposal, and one advantage of this approach is that it avoids the need to make absolute judgements, which can be very difficult where qualitative assessment is involved. It does however require detailed information on all the alternatives and the ranking by different criteria may require considerable knowledge and experience.

(vi) **Decision analysis:** Decision analysis is a systematic procedure for deciding between various options. Although developed primarily to assist management decisions, it has been used in a slightly modified form to evaluate projects, e.g. possible flood alleviation channel alignments. After objectives for the decision have been clarified and then weighted, a set of possible options are identified and evaluated against these objectives. This should highlight a group of high scoring, tentatively acceptable options. For a planning authority to use the method effectively, it would be necessary for the authority to define its objectives clearly and concisely. The weighting of objectives could be difficult, politically contentious and cause delay, whilst additional data may be required in order to assess the alternative options.

(vii) **Safe Minimum Standards:** The concept of Safe Minimum Standards is based on the proposition that critical resources such as soil, water, plants and animals should be protected at some minimum level as a buffer against both unexpected environmental behaviour and the consequent risk of irreversible change. A recent report for the Countryside Commission (Barker, A. et al 1993) identified two principal types of environmental standard: pollution discharge standards such as statutory water quality objectives; and `stock' standards, ie. standards which ascribe a value to specific geographical areas (eg. SSSIs, Scheduled Ancient Monuments, etc.). Whilst not, in itself, a methodology for making planning decisions, the concept of Safe Minimum Standards can provide a useful 'first sieve' in the consideration of development proposals. The use of this concept by planning authorities is, however, severely limited at present by the fact that minimum standards for many indicators of environmental quality have yet to be set.

5.19 The decision-making methods outlined above all draw on monetary evaluation techniques in seeking to present environmental costs and benefits alongside those of a purely economic nature, although as **Appendix 4** explains, inaccurate or poor quality information on environmental impacts can hinder the successful translation of environmental impacts into monetary values. The principal monetary evaluation techniques are:

(i) **The Contingent Valuation Technique:** This technique obtains evaluations by using surveys to ask people directly what they ar (WTP) for an environmental benefit, or what they are willing to a an environmental loss.

(ii) **The Travel Cost Technique:** The Travel Cost technique uses the incurred by individuals travelling to a site as a proxy for its recreation value. For example, it would be assumed that each visitor to a country park places a value on the park which is at least equivalent to the costs incurred to make the visit (transport, entrance fees, opportunity cost of time, etc.).

(iii) **Hedonic Pricing:** This technique uses information revealed by purchasing decisions to estimate the monetary value of environmental goods which do not have market prices. House prices are most commonly used for this. Because these vary in accordance with environmental quality, they can be used to estimate the monetary value of environmental attributes such as landscape amenity, noise and air quality.

(iv) **The Shadow-Project Approach:** This pricing approach examines the financial costs of providing an equal alternative environmental good elsewhere. Taking an area of marshland as an example, various alternatives can be considered: asset reconstruction (providing an equal alternative site elsewhere); asset transplantation (moving the existing habitat to a new site); and asset restoration (enhancing an existing degraded habitat elsewhere).

(v) **The Dose-Response Technique:** The dose-response technique is based upon establishing a relationship between a specific amount of pollution and any observed damage it causes. Once a dose-response relationship has been established, the economic costs associated with a specified amount of damage (or a unit of damage) can be calculated. This approach has been used for air pollution in connection with respiratory illness, death, building damage and harm to vegetation.

5.20 In **Appendix 5**, the case of a proposed city airport has been taken as an example of a planning decision involving an ES, and used to test the possible applicability of monetary evaluation techniques. The study examines the potential applicability of techniques, such as contingent valuation and hedonic pricing, to the evaluation of constructional impacts (loss of ancient woodland, part of a golf course and green belt land) and 'airport use related impacts' (largely air and noise pollution). It concludes that in this example, the Travel Cost Method (using costs incurred by visitors) could apply to impacts on the golf course and green belt areas, whilst Hedonic Pricing (using property prices as indicators) would be most applicable to the air and noise impacts. Contingent Valuation (using survey information of peoples' willingness to pay for benefits or accept compensation for costs) is considered to be potentially applicable to all impacts studied.

VIEWS OF LOCAL AUTHORITY PLANNING OFFICERS AND OTHER PRACTITIONERS

5.21 Some resistance was met from local authority planning officers and other practitioners who were interviewed as part of the research to the ideas canvassed for introducing formalised methods and techniques to help evaluate environmental information in the context of EA. This is borne out by evidence from practice. For example, only two of the authorities interviewed on the telephone were able to point to examples of the practical application of monetary valuation techniques.

5.22 One of the environmental consultants described how the contingent valuation method had been used to calculate recreation costs and benefits as an input to a feasibility study for flood defence works. However, the consultant was of the view that whilst it was an interesting piece of work, it did not greatly enhance the evaluation process.

5.23 The response was much the same when referring to decision-making methods. With the exception of one or two road proposals where pair-wise comparison was used to appraise alignment options, the planning officers interviewed were firmly of the view that planning recommendations are a matter for professional judgement and it is unlikely that there is a method or technique sophisticated enough to replace this.

5.24 Moreover, many interviewees felt that the whole idea of ascribing a monetary value to an environmental good is fundamentally flawed. This is not to say that they did not see merit in comparing economic, social and environmental factors within a common framework, but rather that they believed it should be accepted that there will always be intangibles in such an equation which rely on professional and political considerations for resolution.

5.25 At a more practical level, all local authority planning officers expressed concern about the resource implications if formal methods and techniques, especially those requiring monetary evaluation, were to become standard practice. Not only did these comments refer to the additional cost of adding another layer of assessment to a system which they consider is already under severe strain, but also the fact that, in their view, most planners do not possess the requisite skills[5] .

THE APPLICABILITY OF METHODS AND TECHNIQUES OF EVALUATION

5.26 The technical shortcomings associated with many of the evaluation methods and techniques considered, coupled with the general resistance to their use in practice expressed by the practitioners we interviewed, suggests that evaluation methods and techniques are unlikely to offer much assistance to local authority planning officers dealing with ESs at the present time. The one possible exception appears to be where the decision is taken at the outset of the formulation of a development proposal to make use of a particular method or technique. If this occurs at an early enough stage it may be possible to ensure that the information collected is directly relevant to the method or technique being employed. This view appears to be supported by what was a consistent message from many of the people interviewed during the research, ie. that the earlier a planning authority is involved in the formulation of development proposals, the more likely that it will be able to ensure that the environmental information collected relates directly to the assessment of likely impact.

5.27 All methods and techniques for dealing with evaluation require a sound information base, and some also require carefully researched public perception surveys. Perception surveys should be based on a clear understanding of the development options, and this may mean that the work needs to be programmed into the EA itself. Provided such action is taken, there are a number of techniques like contingent

---

[5] A survey of ten of the major university schools or departments offering either undergraduate or postgraduate courses in planning or EA undertaken as part of the research, revealed that no course units were specifically time-tabled for either the integration of environmental information into planning decisions or actual decision-making techniques (although such topics may be touched upon in other components of the courses, for example, in units on environmental planning. Only two course prospectuses made reference to the subject of decision-making at all, with one of these providing a unit on `applying elements of sociological theory to the analysis of planning decisions and their outcomes in practice', and offering planning students the opportunity to observe decision-making in local planning committees.

valuation (see **Appendix 4**) which can help to clarify public responses to particular elements of a project.

5.28 The case studies did not yield any examples of the methods and techniques being used to evaluate environmental information. However, the discussions with practitioners did serve to identify one or two projects where an attempt had been made to employ a particular decision-making method to assist with the formulation and assessment of development options. In both cases these were flood defence schemes promoted by a public agency (the National Rivers Authority) and falling outside planning legislation. This in itself appears to be an important research finding. Given that evaluation methods and techniques are only likely to offer much assistance to the decision-taking authority if they are specifically built into the early stages of the formulation of a project, it may be that they will only really be embraced as part of a public sector development project.

5.29 Related to the comments in paragraph 5.26 above, is the conclusion that certain types of project are more amenable to the use of formalised evaluation methods and techniques than others. There are examples of projects with genuine alternatives, like different road alignments, or a range of sites for waste disposal by landfill, where evaluation methods and techniques have been used to reach informed judgements on preferred sites. For the most part, these are projects promoted by public sector agencies. The method of pair-wise comparison of options, in particular, is in relatively common use, especially for the evaluation of options for linear projects.

5.30 None of the case study projects promoted by private sector applicants had a full comparison of options, including the do-nothing option, in the ES. This is not to say that some form of site selection did not take place. It clearly did. However, the criteria were more to do with commercial viability than environmental impact. For obvious reasons, this information was not presented in the environmental statement.

WEIGHING-UP INFORMATION TO REACH A DECISION

5.31 The resistance to evaluation methods and techniques referred to in paragraph 5.21 above stems from the view that planning officers' recommendations are a matter for professional judgement, expressed as recommendations on planning applications, and it is unlikely that there will ever be a method or technique sophisticated enough to deliver the recommendation itself. Nevertheless, information generated by quantitative and qualitative methods and techniques may be available when considering the likely effects of development proposals and whether the proposals should be approved. A local authority planning officer has to draw together the often disparate output from these, take account of factors which are not susceptible to them and give weight to the significance of each. It is the exercise of professional judgement in this process which leads planners away from an over-reliance on evaluation methods and techniques and such reliance could only be justified to the extent that the methods and techniques reduce the extent of uncertainty confronting planners.

5.32 The most important factor is the planning policy context set out in the statutory development plan. All development control decisions are required to accord with the development plan unless material considerations indicate otherwise. Although a number of the case studies pre-dated the introduction of this provision[6] , established planning policy was the key consideration in virtually every case. This is best illustrated by considering how the issues were presented to planning committees.

5.33 The particular approach varied from authority to authority and depended upon:

---

[6] Planning and Compensation Act 1991.

- the established methods of presentation for planning applications generally;
- the experience of the authority in dealing with applications accompanied by an ES;
- the sensitivity of the application and the main environmental issues.

5.34 The ES was usually included in the Background Papers available to members, although one planning authority included the text of the ES as an appendix to the Committee Report. Another authority simply reproduced the non-technical summary extracted from the ES. All the Committee Reports comprised the following elements:

- description of the proposals;
- planning history/background;
- planning policy context;
- summaries of consultations, representations and objections;
- officer's review of the issues raised by the proposal;
- officer's recommendation.

5.35 The Committee Reports generally made use of the environmental information presented in the ES in two ways. Firstly, the information was used to make statements of fact about potential environmental impacts, for example the loss of habitat caused by a development proposal. Secondly it was used to make an appraisal of the policy implications of a project. Again using the loss of habitat example, the environmental information would be used by the local authority planning officer to determine the planning policy implications of the impacts identified. In almost all the case studies most emphasis was placed in the officer's report to Committee on the extent to which the proposal was consistent with, or contrary to, established planning policy, although it sometimes proved difficult to appreciate the relative weight placed on particular national planning guidance or the policies in development plans which may be relevant. The use of simple matrices depicting the type and magnitude of potential impacts could go a long way towards overcoming such difficulties.

5.36 Another aspect of the process of weighing-up environmental information highlighted as an area of concern by some local authority planning officers was the extent to which environmental issues do or do not fall under planning control. In certain cases, the view was expressed that some of the potential environmental impacts caused by projects which should ideally be dealt at the planning application stage, do not in practice get dealt with until a subsequent stage (usually the granting of the necessary discharge/abstraction consents or licences). In other cases, the contrary view was expressed, ie. that the planning application often covers issues going beyond planning. Both of these views reflect the fact that the system of regulatory control in the UK is such that a planning authority's consideration of a project is directed to its planning merits, leaving consideration of some other environmental issues to other bodies.

6.0 CONCLUSIONS AND AREAS WHERE GUIDANCE MAY BE HELPFUL

6.1 The research reported in this document was commissioned as the first stage in the production of a good practice guide to assist local authority decision-makers in their evaluation of environmental information. This is information which must be taken into account when determining planning applications which are the subject of an environmental statement. The draft guidance itself is the subject of a separate document. The purpose of the research was to review existing literature and planning authority experience, including the use of monetary and non-monetary evaluation techniques, and to determine the nature and extent of guidance required to assist local authority planning officers, planning committees and others to:

(i) consider whether submitted ESs are adequate or whether additional information needs to sought from the applicants; and

(ii) evaluate the information in the ES and any representations thereon from statutory consultees and others, so that this can contribute to an informed decision on the application for planning permission.

6.2 The main findings of the research broadly fall under three headings:

- the procedures used by planning authorities to handle environmental statements and evaluate information;
- the quality and adequacy of environmental statements;
- the methods and techniques of evaluation and their potential use.

The conclusions in each case are set out below.

THE PROCEDURES USED BY PLANNING AUTHORITIES TO HANDLE ENVIRONMENTAL STATEMENTS AND EVALUATE INFORMATION

6.3 Not surprisingly, the type and effectiveness of procedures used by planning authorities to handle ESs are strongly influenced by the number of statements that the authority has had to deal with. Most of the authorities which have dealt with a relatively large number of ESs have developed effective procedures for steering them through the development control process based on the following stages:

(i) Pre-application discussions/scoping.

(ii) Review of the content and quality of an ES.

(iii) Consultation with statutory and non-statutory consultees.

(iv) Public consultation.

(v) Evaluation of the environmental information.

(vi) Presentation of information to decision-makers.

6.4 However, it also seems to be the case that some planning authorities are ill-prepared to deal with an ES when it is first submitted. Junior officers who are often the first to examine the planning application and ES are sometimes unaware of the requirements of the Regulations, and unsure of how to proceed with examining and processing the ES, let alone evaluate its contents. The causes of this uncertainty appear to reflect a combination of a lack of time available fully to assess an ES, lack of experience, lack of in-house specialists and inadequate resources to `buy-in' specialist help.

6.5 Discussions with local authority officers have indicated that the overall weight and style of presentation of an ES can affect the way in which it is processed. Very lengthy and apparently competent ESs may receive less critical scrutiny than brief and poorly presented documents even though the actual content of the former may be no more reliable. This reflects the fact the authorities are sometimes unwilling or unable to carry out a critical review of specialised areas of work in which officers have little knowledge or experience. Indeed it appears from some of the case studies that inadequate ESs have been accepted, processed and used as a basis for decisions.

6.6 We are therefore of the view that many planning authorities, notably those who have relatively little experience of dealing with ESs, would benefit from guidance which follows the development control process through from initial receipt of the planning application and ES to final decision. The aim should be to suggest ways in which 'environmental information' can be checked and processed to meet the requirements of the Regulations, evaluated by the planning officer and presented to the planning committee.

THE QUALITY AND ADEQUACY OF ENVIRONMENTAL STATEMENTS

6.7 Our research has shown that most of the literature on methods for reviewing ESs tends to concentrate on the quality of the written statement without necessarily exposing the adequacy of the information contained within them. Also, some divergence of opinion exists amongst practitioners from both planning authorities and consultancies about the degree of attention which should be paid to the quality of the ES, as opposed to the adequacy of the information on which the planning decision is eventually founded. We are of the view that both aspects of the evaluation process are important.

6.8 A number of methods have been developed for assessing the quality of an ES, the main examples of which are considered in Section 4 of the research report. Most of them share a common characteristic in that they are derived from the Lee-Colley Review Package in some form. In practice this means that each method establishes a checklist which can be used to ensure that the ES includes all the relevant material. Our view is that such methods, including Lee-Colley, are likely to be most useful in helping a planning authority decide whether the information submitted is adequate or whether further information should be required from the developer.

6.9 Turning to the question of the adequacy of the environmental information within an ES, the EC Directive and the EA Regulations require developers to provide, as a minimum requirement, certain basic levels of information about a project and its likely environmental effects. Developers should therefore be expected to produce ESs of a high standard from the outset. However, it is clear from the research that planning authorities are sometimes obliged to spend time and resources they can ill afford seeking clarification and reworking inadequate information. We therefore recommend that the Regulations should be modified to allow planning authorities to decline to register an Environmental Statement if it contains obvious shortcomings which fail to meet the basic requirements.

6.10 Even when an EA has been well researched and a comprehensive ES has been presented there may often be a prolonged period of negotiation between the developer and planning authority, resulting in amendments to the proposals, and the generation of large amounts of supplementary information, some of which will be provided in documents other than the ES. In these circumstances the question arises as to whether or not a revised and updated ES should be produced on completion of the planning officer's appraisal, either by the developer, or by the authority itself. If by the planning authority, such a document might consist of the factual sections of the report to committee, summarising all the information and views gathered during the course of the evaluation.

6.11 Our research has led us to conclude that the ideal situation would be a combination of the suggestions made above on assessing quality and adequacy. The main objective of the EA system should be to achieve the best possible ES at the outset, and the guidance should set out a cost effective and speedy way of coming to an initial view that this is indeed so. It should also, however, develop a step-by-step approach for dealing with incomplete or inadequate information. The Regulations empower a planning authority to require the developer to provide specified further information, and the planning authority shall not determine the application until the further information is provided.

METHODS AND TECHNIQUES OF EVALUATION AND THEIR POTENTIAL USE

6.12 The questions which lie at the heart of this research are concerned with the ways in which the wide range of environmental information generated by EA should be sorted, prioritised, organised and presented; and whether there are methods and techniques which can be used to identify the key issues and assist the decision-maker in reaching a balanced conclusion. Our research suggests that formal methods and techniques are not, in general, used by planning authorities to evaluate environmental information where projects are subject to EA. That is not to say that such methods and techniques do not have a role to play and, although no one approach should be regarded as a panacea, there are a number which might be employed to clarify specific areas of uncertainty. The usefulness of these is likely to be enhanced if they are initiated at an early stage of an EA study; if they are easily understood and simple to use; and if they do not involve substantial time and expense.

6.13 The impetus to develop some convenient framework for decision-making has come from three main areas of practical and academic research: engineering; economics; and planning, design and management. The research has considered the contributions from each of the principal disciplines, and looked in detail at seven evaluation and decision-making methods. For the purpose of this report a distinction is made between 'techniques' which are used in specialist fields to evaluate the importance of individual environmental, social and economic topics, and 'evaluation and decision-making methods' which seek to aggregate information from all sources and draw conclusions about their relative importance. All seven methods draw on various monetary evaluation techniques.

6.14 The research revealed that the methods and techniques of evaluation currently available all appear to suffer from technical shortcomings, as well as being the subject of a fair degree of scepticism amongst local authority planning officers about their practicability. This is shared by our research team. These two factors have led us to the conclusion that, regardless of the merits or limitations of particular methods and techniques, it would be unrealistic and may be undesirable to expect to change the approach of local authority planning officers in the short term.

6.15 The choice of methods and techniques for any kind of analysis ultimately depends on the nature of the problem to be addressed. Planning authorities have to make judgements about large numbers of planning applications; and the background against which these judgements have to be made is the development plan for the area. Generally, the commercial viability of development is not a material consideration for the purposes of planning. Nevertheless, all planning decisions require that the potential adverse and beneficial effects on the natural and man-made environment should be carefully considered.

6.16 The purpose of an ES is to provide information to enable the planning authority to determine the extent to which a particular project is likely to give rise to environmental concerns because of its size or location or because of the nature of the processes involved, and the extent to which those concerns can be mitigated. Ideally this information should be presented in the ES in a form which quantifies the various positive and negative environmental impacts. This is where the techniques of monetary valuation have relevance. However, the application of these techniques can be difficult, expensive and time consuming. Very often the most that can reasonably be expected is that an ES will assess the extent of the various different kinds of environmental impact in physical and not monetary terms. In the evaluation (within the context of the development plan) of environmental information and other material considerations, there will remain a considerable onus on the professional judgement of local authority planning officers.

6.17 We are of the opinion, therefore, that what would be useful in guidance is a step-by-step analysis of the evaluation process as practised by the most experienced authorities, and provision of worked examples which illustrate good practice where it exists. The research, notably the views expressed by local authority planning officers, indicates the guidance should focus on some of the simpler methods which exist for the evaluation of environmental information and presenting the conclusions to a planning committee. The guidance could go on to illustrate how simple methods can be expanded to incorporate more advanced types of analysis where this may be appropriate.

APPENDIX 1 : BIBLIOGRAPHY

BIBLIOGRAPHY

1: GENERAL

Barker, A., Bowers, J., Hopkinson, P. and Lydall, K. (1993)
Environmental Standards: Issues for the Countryside Commission.

Bisset, R (1984)
Post-development Audits to Investigate the Accuracy of Environmental Impact Predictions. In Zeitschrift fur Umweltpolitik, 7, 463-84.

Bowers, J.K., Bristow, A.L. and Hopkinson, P.G. (1991)
The Treatment of Landscape in Public Investment Appraisal.

Buckley, G.P. ed, (1989)
Biological Habitat Reconstruction, Belhaven Press, London.

Cheshire County Council. (1989)
Planning Practice Notes No. 2. Cheshire Environmental Assessment Handbook. October 1989. Cheshire County Council. No. 404.

Cloke, P.J. (1983)
An Introduction to Rural Settlement Planning, Methuen, London and New York.

Corkindale, J. (1993)
Environmental Impact Assessment, Environmental Appraisal and Decision Making. EIA Conference. University of Manchester, 1993.

Department of the Environment/Welsh Office. (1989)
Environmental Assessment. A Guide to Procedures. HMSO, London.

Department of Transport (1983)
Manual of Environmental Appraisal. HMSO, London.

Environmental Assessment Panel. (1976)
Guidelines for Preparing Initial Environmental Evaluation, Environmental Assessment Panel, Canada.

Essex Planning Officers Association. (1992)
The Essex Guide to Environmental Assessment. Essex County Council.

Geraghty, P.J. (1992)
Environmental Assessment and the Application of an Expert Systems Approach. Town Planning Review, Vol 63, No. 2.

Goodland, R. (1989)
The Environmental Implications of Major Projects in Third World Development. In Morris, P. (Ed). Major Projects and the Environment. Major Projects Association, Oxford.

Harrison, A.J. (1977)
Economic and Land Use Planning. Croom Helm, 1977.

Kent County Council (undated).
Kent Environmental Assessment Handbook, Planning Department, Kent County Council.

Lave, L. and Seskin, E. (1977)
Air Pollution and Human Health. John Hopkins University Press.

Lee, N. (1993)
The Quality of Environmental Statements in the UK, in Integrated Environmental Management, March 1993.

Lee, N. and Brown, D. (1992)
Quality Control in Environmental Assessment. Project Appraisal, Vol 7, No. 1.

Lee, N (Ed). (1992)
Special Issue on Strategic Environmental Assessment. In Project Appraisal, Vol 7, No. 3.

Lee, N. and Dancey, R. (1993)
The Quality of Environmental Impact Statements in Ireland and the UK: A Comparative Analysis Project Appraisal, Vol 8, No. 1. Beech Tree Publishing, Guildford.

Munro, D.A., Bryan, T.J. and Matte-Baker, A. (1986)
Learning from Experience: A State-of-the-Art Review and Evaluation of Environmental Impact Assessment Audits. CEARC, Canada.

Muntan, R.J.C. (1981)
Agricultural Land Use in the London Green Belt.

Overseas Development Administration. (1989)
Manual of Environmental Appraisal. HMSO, London.

Robinson, G.M. (1990)
Conflict and Change in the Countryside, Belhaven Press, London.

Rodwell, J. 1986
National Vegetation Classification: Woodlands and Scrub, University of Lancaster, unpublished report to the NCC.

Therival, R., Wilson E., Thompson S., Heaney, D. and Pritchard D. (1992)
Strategic Environmental Assessment. Earthscan, London.

Thomas, R.S., Winfield, P.G., Brooke, J.S., Pearce, D.W., Turner, R., Newbold, C. 1991
Consultants Report on Evaluation of Guidance and Selected Coastal Defence Schemes. Prepared for the National Audit Office.

Turner, R.K. and Brooke, J.S. 1990
The Development of an Environmental Assessment Manual - A Pilot Study. Report for the Ministry of Agriculture, Fisheries and Food, by the Environmental Appraisal Group, University of East Anglia, Norwich, NR4 7TJ.

Wathern, P. (Ed). (1990)
Environmental Impact Assessment. Theory and Practice. Unwin Hyman, London.

Wood, C. and Jones, C. (1991)
Monitoring Environmental Assessment and Planning Report prepared for the Department of the Environment, EIA Centre, University of Manchester/HMSO.

Wood, C., Lee, N. and Jones, C.E. (1991)
Environmental Statements in the UK. The Initial Experience. Project Appraisal Vol 6, No. 4 . Beech Tree Publishing, Guildford.

World Bank. (1991)
Environmental Assessment Source Book Volume II. Sectional Guidelines. Technical Paper No.140. World Bank, Washington D.C., USA.

World Bank. (1991)
Environmental Assessment Source Book Volume III. Guidelines for Environmental Assessment of Energy and Industry Projects. Technical Paper No. 154. World Bank, Washington D.C. USA.

2: ADEQUACY OF ENVIRONMENTAL STATEMENTS

Council for the Protection of Rural England. (1990)
Environmental Statements: Getting them Right. CPRE, London.

Countryside Commission. (1991)
Environmental Assessment, the Treatment of Landscape and Countryside Recreation Issues. Countryside Commission Publications, Manchester.

Elkin, T.J. and Smith, P.G.R. (1988)
What is a good Environmental Statement? Reviewing Screening Reports from Canada's National Parks. Journal of Environmental Management, 26, 71-89. Academic Press.

English Nature, The Countryside Council for Wales and Scottish Natural Heritage. (1992)
Environmental Assessment Handbook.

Institute of Environmental Assessment (1991)
Registration Procedures and Review Criteria.

Institute of Environmental Assessment. (1993)
Guidelines to the Ecological Input to Environmental Assessments in the UK (draft).

Lee, N. and Colley, R. (1990)
Reviewing the Quality of Environmental Statements, EIA Centre, University of Manchester. Occasional Paper No. 24.

Land Use Consultants. (1985)
Channel Fixed Link, Environmental Appraisal of Alternative Proposals. Land Use Consultants for the Department of Transport, HMSO, London.

Ross, W.A. (1987)
Evaluating Environmental Impact Statements, Journal of Environmental Management, 25, 137-147.

Street, E. (1993)
Notes from the coal-face on Environmental Assessment. Planning, 11th June 1993

Tomlinson, P. and Bisset, R. (1983)
"Environmental Impact Assessment, Monitoring and Post-Development Audits", in PADC (eds), Environmental Impact Assessments, Martinus Nijhoff, The Hague, 405-425.

Tomlinson, P. (1989)
Environmental Statements: Guidance for Review and Audit. The Planner, 3rd November 1989.

3: EVALUATION METHODS AND TECHNIQUES

a) General

Key Texts

Barde, J. and Pearce, D.W. (eds). (1991)
Valuing the Environment: Six Case Studies. Earthscan Publications Ltd., London.

This book looks at six countries where monetary evaluation of environmental costs and benefit techniques have been applied. The case studies, written by leading experts in each nation, show how these methods are being used in the UK, Norway and Italy, and the ways in which they are already extensively used in the USA, Germany and the Netherlands. The authors describe how significant advances are being made in this field but also highlight the obstacles to the use of the techniques: the lack of knowledge of environmental economics at Government level; the competition from other Government priorities; and the failure of environmental groups to grasp the importance of financial evaluation to their cause.

Department of the Environment. (1992)
Policy Appraisal and the Environment. HMSO, London.

This guide was written for Civil Service administrators, many of whom are involved in policies which have significant effects on the environment. It considers, amongst other things, the process of policy appraisal; the nature of environmental impacts; gathering information on and handling environmental impacts; and the costs and benefits of different policy options. The appendices highlight methods of valuation and identification of impacts and some case studies.

Department of Transport. (1992a)
Assessing the Environmental Impact of Road Schemes. The Standing Advisory Committee on Trunk Road Assessment (SACTRA). HMSO, London.
This report is divided into six parts. Part one gives the background to the report. Part two describes the manner in which the planning of trunk road schemes evolves and passes through various stages of assessment and approval. Part three describes the development of the appraisal methods used, gives a summary of the views of SACTRA and ACTRA in earlier reports and gives an account of the EC requirements for EA. Part four discusses the adequacy of current appraisal methods and gives recommendations for their improvement. Part five considers the extent to which a greater degree of monetary valuation of environmental effects is either feasible or desirable, and recommendations are made about the further development of the economic assessment of such effects. Finally part six gives some conclusions and general recommendations.

Department of Transport. (1992b)
Assessing the Environmental Impact of Road Schemes. Response by the Department of Transport to the report by the SACTRA. HMSO, London.

This report is a response to the SACTRA report (see above). It gives details of the Department of Transports methods of assessment of environmental impacts such as their timing, content and presentation. Some comment is made on traffic forecasting and COBA, the specialised computer program for economic assessment.

Hanley, N. (1990)
Valuation of Environmental Effects. Final Report - Stage One.

This paper marks the end of the first stage of a research project to assess whether the various techniques available to value non-market environmental effects are applicable in the context of Scottish Office and Scottish Enterprise interests. Firstly, it considers the idea of economic value as it applies to the environment and secondly describes and explains five methods proposed by economists for valuing environmental factors. These methods are: Contingent Valuation, Contingent Ranking, the Travel Cost Technique, Hedonic Pricing and the Avoided Cost Approach. The report concludes by considering the relative merits of the valuation methods discussed, how they might be used in practice, and how easily they can be incorporated within overall investment appraisal schemes.

Hanley, N. (1992)
Valuation of Environmental Effects. Final Report - Stage Two

This is the second stage of the research into the applicability of non-market environmental valuation techniques in the context of Scottish Office and Scottish Enterprise interests. It tests the applicability of some of the most promising techniques by means of a series of case studies, selected so as to represent reasonably typical examples of projects undertaken by the Scottish Office and Scottish Enterprise.

Hanley, N. (1992)
Valuation of Environmental Effects - Appendices.

This part of the report describes and discusses the case studies used to test the valuation techniques discussed in the first and second stages of the report (see above).

Rendel Planning and Environmental Appraisal Group (UEA). (1992)
Environmental Appraisal: A Review of Monetary Evaluation and Other Techniques. Dept. of Transport.

This report considers both monetary and non-monetary evaluation techniques in the context of the environmental appraisal of trunk road schemes in the UK. It discusses the advantages, disadvantages and applicability of these techniques. Also included is a review of practice overseas. The report concludes with a summary and recommendations.

<u>Other Relevant Texts</u>

CNS Scientific and Engineering Services (undated)
Economic Value of Improvements to the Water Environment.

Dixon, J.A., Carpenter, R.A., Fallon, L.R., Sherman, P.B. and Manipomoke, S. (1988)
Economic Analysis of the Environmental Impacts of Development Projects. Earthscan Publications Ltd. London.

Environmental Protection Agency. (1985)
Methods Development for Environmental Control. Benefits Assessment. Volumes I to X. Washington.

Goodland, R. and Ledec, G. (1987)
Neoclassical economics and principles of sustainable development. Ecological Modelling, 38, 19-46.

Gray, R. (1993)
Accounting for the Environment. Paul Chapman Publishing Ltd, London.

H.M. Treasury. (1977)
Policy Evaluation: A Guide for Managers. HMSO, London

H.M. Treasury. (1977)
Economic Information for Environmental (Anti-Pollution) Policy. HMSO, London.

H.M. Treasury. (1991)
Economic Appraisal in Central Government Departments. HMSO, London.

Hanley, N. and Knight, J. (1992)
Valuing the Environment: Recent UK Experience and an Application to Green Belt Land. Journal of Environmental Planning and Management, Vol. 35, No. 2.

Jacobs, M. (1991)
The Green Economy: Environment, Sustainable Development and the Politics of the Future. Pluto Press, London.

Pearce, D.W. and Markandya, A. (1989)
Environmental Policy Benefits: Monetary Valuation. OECD.

Smith, V.K. and Desvouges, W. (1986)

Measuring Water Quality Benefits, Boston, Kluwer-Nijhoff.

Willis, K.G. and Benson, J.F. (1988).
Valuation of Wildlife. In Turner, R.K. (ed). Sustainable Environmental Management, Belhaven Press, London.

Willis, K. and Whitby (1985).
In Journal of Rural Studies, 1, 147-162.

World Bank. (1991)
Environmental Assessment Source Book, V5ol. 1: Policies, Procedures and Cross-sectoral Issues. Technical Paper 139. World Bank, Washington D.C., USA.

b) Monetary

Key Texts

Pearce, D.W., Markandya, A. and Barbier, E.B. (1989)
Blueprint for a Green Economy. Earthscan Publications Ltd. London

Originally prepared as a report for the Department of the Environment, this book presents a series of practical proposals for financing a sustainable environment. The book covers discussions on the meaning of sustainable development, valuing the environment, accounting for the environment, project appraisal, discounting the future, and prices and incentives for environmental improvements.

Pearce, D.W. and Turner, R.K. (1990)
Economics of Natural Resources and the Environment. Harvester Wheatsheaf.

This book aims to provide a thorough grounding in the economics required to understand national, international and global environmental problems. It deals fully with the orthodox theories of the economics of pollution and optimal depletion rates for natural resources. Environmental ethics, pollution control policy and sustainable development are also discussed.

Winpenny, J.T. (1991)
Values for the Environment: A Guide to Economic Appraisal. HMSO London.

This practical guide to the economic treatment of the environment in project appraisal uses cost-benefit analysis as the decision framework. The main environmental impacts of projects and the methods available for placing economic values on them are discussed. The feasibility of environmental valuation is then illustrated for projects in a variety of sectors, including natural resources, infrastructure, the built environment, industry, mining and tourism. A review of relevant policy issues concludes the guide.

Other Relevant Texts

Anderson, G. and Bishop, R. (1986)
The Valuation Problem, in Bromley, D. (ed), Natural Resource Economics, Boston: Kluwer-Nijhoff.

Arrow, K., Solow, R., Portney, P.R., Leamer, E.E., Radner, R. and Schuman. (1983)
Report of the NOAA Panel on Contingent Valuation, Report to the General Counsel of the US National Oceanic and Atmospheric Administration, Resources for the Future, Washington, D.C.

Bateman, I.J. and Turner, R.K. (1992)
Evaluation of the Environment. The Contingent Valuation Method. CSERGE Working Paper 92-18, University of East Anglia, Norwich NR4 7TJ.

Bateman, I.J. (1992)
The Economic Evaluation of Environmental Goods and Services in Integrated Environmental Management, No.14, 1992, pp.11-14.

Bateman, I.J., Langford, I.H., Willis, K.G., Turner, R.K. and Garrod, G.D., (1993).
The Impacts of Changing Willingness to Pay Question Format in CVM Studies: An Analysis of Open-Ended, Iterative Bidding and Dichotomous Choice Formats.

CSERGE Working Paper GEC 93-05, University of East Anglia, Norwich.

Benson, J.F, and Willis, F.G. (1991)
Valuing informal recreation on the forestry commission estate. Forestry Bulletin 104, HMSO, London.

Bishop, R.C. (1978)
Endangered Species and Uncertainty: The economics of a Safe Minimum Standard, American Jnl, of Agricultural Economics, 60, 10-18.

Bishop, R. (1982)
Option Value: An Exposition and Extension in Land Economics, February 1982.

Bishop, K. (1992)
Assessing the Benefits of Community Forests: An Evaluation of the Recreational User Benefits of Two Urban Fringe Woodlands. Journal of Environmental Planning and Management, Vol. 35, No. 1.

Bowers, J. (1993)
Pricing the Environment: A Critique in International Review Applied Economics, Vol. 7, part 1, pp. 91-107.

Brookshire, D.S., Thayer, M.A., Schulze, W.D., and D'Arge, R.C. (1982)
Valuing Public Goods: A Comparison of Survey and Hedonic Approaches. The American Economic Review, March 1982.

Ciriacy-Wantrup, S.V. (1986)
Resource Conservation: Economics and Policies, Berkeley and Los Angeles: University of California.

Costana, R., Faber, S.C. and Maxwell, J. (1989)
Valuation and Management of Wetland Ecosystems, Ecological Economics, 1 (4).

Crocker, T.D. (1986)
On the Value of the Condition of a Forest Stock, in Land Economics 61(3) pp. 244-254.

Department of the Environment. (1989)
Handbook of Estate Improvements: Part 1, Appraising Options, HMSO, London. ISBN 0-11-752232-5.

Desvousges, W.H. et al (1987)
Option Price Estimates for Water Quality Improvement, Journal of Environmental Economics and Management (14) pp. 248-267.

Desvousges, W.H., Smith, V.K. and McGivney, M.P. (1983)
A Comparison of Alternative Approaches for Estimating Recreation and Related Benefits of Water Quality Improvements. EPA, Washington.

Dixon, J.A. and Sherman, P.B. (1991)
Economics of Protected Areas, A New Look at Benefits and Costs. Earthscan Publications Ltd, London. ISBN 1-85383-097-6.

ESRC (1990-1992)
Countryside Change Initiative Working Paper Services. Various Papers including Working Papers 5, 7, 10, 12, 13, 14, 19, 21, 25 and 28.

Freeman, A.M. (1979)
Hedonic Prices, Property Values and Measuring Environmental Benefits. In Scandinavian Journal of Economics. Vol. 81, 154-173.

Freeman, A.M. (1982)
Air and Water Pollution Control: A Benefit-Cost Assessment. John Wiley & Sons, Inc. USA. .

Garrod, G.D, and Willis, K.G. (1992)
The Environmental Economic Impact of Woodland as Two Stage Hedonic Price Model of the Amenity Value of Forestry in Britain. Applied Economics, 24, 715-728.

Grosslink, J.G., Odum, E.P. and Pope, R.M. (1974)
The Value of the Tidal Marsh, Centre for Wetland Resources, Louisiana State University.

H.M. Treasury. (1977)
Economic Information for Environmental Anti-Pollution Policy. Government Economic Service Occasional Papers. HMSO, London.

Hanley, N.A. et al. (1991)
Environment Economics and Sustainable Development in Nature Conservation. Report to the Nature Conservancy Council.

Harris, B.S. (1984)
Contingent Valuation of Water Pollution Control, in Journal of Environmental Management, 19, pp. 199-208.

Helm, D. and Pearce, D.W. (Eds), (1990)
Economic Policy Towards the Environment in Oxford Review of Economic Policy, Vol. 6, No. 1.

Hopkins, J. (1988)
Transplantation Guidelines, NCC, Peterborough, unpublished.

Hufschmidt, M.M., James, D.E., Meister, A.D., Bowers, B.T., and Dixon, J.A. (1983)
Environment, Natural Systems and Development: An Economic Valuation Guide. John Hopkins University Press, London.

Johansson, P.O. (1987)
The Economic Theory and Measurement of Environmental Benefits. Cambridge University Press.

Khan, J.R. and Kemp, W.M. (1985)
Economic Losses Associated with the Degradation of an Ecosystem: The case of the submerged aquatic ecosystem in Chesapeake Bay. In Journal of Environmental Economics and Management. 12, (3).

Krutilla, J. (1967)
Conservation Reconsidered, American Economic Review, 47, 777-786.

Lichfield, N. (1969)
Cost Benefit Analysis in Urban Expansion. A Case Study: Peterborough. Regional Studies Vol. 3.

Lichfield, N. (1988)
Economics of Urban Conservation. Cambridge University Press, Cambridge.

Ministry of Agriculture, Fisheries and Food. (1993)
Flood and Coastal Defence. Project Appraisal Guidance Notes.

Ostro, B. (1983)
The Effects of Air Pollution on Work Loss and Morbidity. Journal of Environmental Economics and Management, Vol. 10, 371-382.

Pearce, D.W. and Turner, R.K. (1990)
Economics of Natural Resources and the Environment. Harvester Wheatsheaf. ISBN 0-471-08985-0.

Pearce, D.W. (Ed). (1991)
Blueprint 2: Greening the World Economy. Earthscan Publications Ltd, London. ISBN 1-85383-076-3.

Schulze et al (1983)
The Economic Benefits of Preserving Visibility in the National Parklands of the South West, in Natural Resources Journal, 23, 149-73.

Sorg, C.F. et al (1986)
Net Economic Value of Cold and Warm Water Fishing in Idaho, in Resource Bulletin RM-12, Rocky Mountain Forest and Range Experimentation Station, US Forest Service, Fort Collins, Colorado.

Turner, R.K. and Bateman, I.J. (1991)
A Critical Review of Monetary Assessment Methods and Techniques. Environmental Appraisal Group, University of East Anglia, Norwich, NR4 7TJ.

Turner, R.K., Bateman. I.J. and Pearce, D.W. (1992)
Valuing Environmental Preferences: The UK Experience. Centre for Social and Economic Research on the Global Environment (CSERGE), Working Paper GEC 92-04, University of East Anglia, Norwich, NR4 7TJ.

Walsh, G.R., Loomis, J.B. and Gillman, R.A. (1984)
Valuing Option, Existence and Bequest Demand for Wilderness, in Land Economics, 60, (i), 15-29.

Ward, W.A. and Deren, J.B. with D'Silva, H.E. (1991)
The Economics of Project Analysis: A Practitioner's Guide. The International Bank for Reconstruction and Development/The World Bank, Washington, USA.

Willis, F.G, and Benson, J.F. (1989)
Recreational Values of Forests. In Forestry, 62,2, 93-100.

Willis, K. et al. (1992)
Urban Development in the Rural Fringe: A Decision Analytical Framework and Case Study of the Newcastle Green Belt, Countryside Change Initiative, Working Paper 31.

Willis, K.G, and Garrod, G.D. (1993)
Valuing Landscape: A Contingent Valuation Approach, Journal of Environmental Management.

c) Non-Monetary

Key Texts

Ciriacy-Wantrup, S.V. (1986)
Resource Conservation: Economics and Policies, Chp. 18). Berkeley and Los Angeles, University of California.

This chapter discusses the concept of 'Safe Minimum Standards' in the context of environmental conservation. Underpinning this concept is the view that critical resources (he cites soil, water, plants and animals) should be protected at some minimum level as a buffer against both unexpected environmental behaviour and the consequent risk of irreversible change.

Hill, M. (1968)
A Goals - Achievement Matrix for Evaluating Alternative Plans.
In Journal of the American Institute of Planners, Vol. 34, pp.19-29

This paper examines the application of some established techniques such as cost-benefit analysis and the planning balance sheet, to plan evaluation. It is argued that these techniques do not fully satisfy the requirements of the rational planning process, and an alternative method of plan evaluation, known as the goals-achievement matrix, is suggested.

Kass, G.S. and Walker, C.E. (1992)
Pairwise: A Tool for Comparative EA. Paper from the International Conference on Environmental Assessment 1992.

This paper describes a methodology that attempts to establish a framework in which comparative environmental assessment can be achieved without recourse to subjective scoring or weighting systems. Pairwise analysis is used for each of the elements of two or more impact matrices representing project alternatives. It is a tool that will aid the identification of a single or a number of preferred options. The methodology is examined using a fictitious case study involving four options for a water supply project.

Kepner, C. and Tregoe, B. (1981)
The New Rational Manager (Chs 4 and 5)

Chapter 4 describes a decision making procedure (Decision Analysis) based on the thinking pattern we all use in making choices. Decision Analysis involves identifying what needs to be done, developing the specific criteria for its accomplished, evaluating the available alternative, relative to those criteria, and identifying the tasks involved. Chapter 5 goes on to describe the uses of Decision Analysis, suggesting that it can be applied to complex decisions and involving both a number of alternatives and "yes/no" decisions that involve only two alternatives. The book is aimed primarily at business managers and contains no examples directly relevant to the Planning Decision.

Lichfield, N. (1970)
Evaluation Methodology of Urban and Regional Plans: A Review. Regional Studies Vol. 4, pp 151-165. Pergamon Press.

This paper describes the generalities of tests of plans and projects, then makes a comparative review of some twenty plan evaluation methodologies. This is done by reference to the criteria to which comprehensive evaluation methodologies should conform. It concludes that the Planning Balance Sheet has greater potential.

Lichfield, N. (1992a)
The Integration of Environmental Assessment into Development Planning: Part 1, some principles. Project Appraisal, Vol. 7, No. 2.

This is the first of two articles considering the need for greater integration of the disciplines of environmental assessment and planning. It considers the principles involved in this integration.

Lichfield, N. (1992b)
The Integration of Environmental Assessment and Development Planning: Part 2. A Case Study. Project Appraisal, Vol. 7, No. 3.

This case study is presented to demonstrate the method described in Part 1 (see above) for including environmental assessment in development planning. It considers the proposed creation of a Headquarters and Business Centre for British Airways at Heathrow Airport. The conclusion is that the benefits accrued from the development out-weigh the disbenefits of the construction period.

Other Relevant Texts

Ashworth, G. (1975)
Environmental Evaluation - A Review of Current Approaches and Methodology. Paper given by Professor G. Ashworth, University of Salford, 26 September 1975.

Chapman, M. (1981)
Decision Analysis. Her Majesty's Stationery Office, London.

Hill, M. (1973)
Planning for Multiple Objectives: An Approach to the Evaluation of Transportation Plans, Regional Science Research Institute, Philadelphia, USA.

Massam, B.H. (1988)
Multi-criteria decision making (MCDM) techniques in planning. In Progress in Planning, Vol. 30, 1, 1-84.

Ratcliffe, J. (1981)
An Introduction to Town and Country Planning. 2nd Edition. Hutchinson.

Voogd, H. (1979)
Multicriteria Evaluation for Urban and Regional Planning.
Technische Hogeschool, Delft.

APPENDIX 2 : SUMMARY OF TELEPHONE INTERVIEWS

SUMMARY OF TELEPHONE INTERVIEWS

1.1 Telephone conversations with 52 planning authorities took place and two responded by way of a letter. The breakdown, by type of authority, is as follows:

Table 1.1 - Type of Authority

Type of Determining Authority	Number Responding
County Councils (England)	10
County Councils (Wales)	1
Regional Councils (Scotland)	2
District Councils (England, Non-Metropolitan)	20
District Councils (England, Metropolitan)	4
District Councils (Wales)	3
District Councils (Scotland)	5
London Borough Councils	5
Development Corporations	3
National Parks	1
	54

1.2 **Number of Environmental Statements Received:** The number of ESs received by each of the authorities who responded is shown in **Table 1.2**. Some officers contacted were unable to give the exact number received, and the figures exclude those ESs where the authority was acting as a consultee only.

Table 1.2 - Number of ESs Received

Number of ESs Received	Number of Authorities
1	10
2	11
3	11
4	6
5	5
6	4
7+	5
not stated	2
	54

Note: Six of the planning officers contacted were either unable or unwilling to give precise details of the number of ESs received. Of these, four officers stated that at least seven ESs had been submitted to their authority, and these are therefore included within this total in Table 1.2.

1.3 **Status of Applications Accompanied by an ES:** Of those ESs for which the authorities were able to give details, the status of each planning application at the time of the telephone survey was as follows:

Table 1.3 - Status of Applications

Status of Application	Number of ESs
Approved	51
Refused (possibly subject to later appeal)	11
Pending	63
Withdrawn	10
Called-in by the Sec. of State	6
At Appeal	5
	146

Note: The total of 146 ESs excludes those relating to the six authorities who were unwilling or unable to give precise details of the ESs they had received.

1.4 It can be seen from **Table 1.3** that more than four times as many applications accompanied by an ES were approved than refused. However, just over half of the applications received were still to be formally determined including those awaiting a planning agreement and those applications subject to appeal or called in by the Secretary of State.

1.5 **Established Procedures:** The discussions with planning officers revealed that, although at first sight there appeared to be two distinct processes in assessing an ES (ie determining the adequacy of the information contained within an ES, and evaluating the information to come to a decision) in practice the two are often indistinguishable. Planning authorities tend to evaluate the information provided and identify deficiencies at the same time (although further information provided in response to the identification of deficiencies also has to be evaluated at some stage).

1.6 Few officers were able to declare that they used formal procedures (eg Lee-Colley) for evaluating an ES as part of a planning application. In the majority of cases, each application was assessed on its own merits, largely based on the issues it raised.

One Borough Council described the steps it goes through to do this as follows :

i) where the authority is advised of the project at the pre-application stage, the applicant is advised in pre-application consultations to follow the procedure set out below:

- carry out a desk study of available environmental information;
- provide a method statement detailing: what further survey work is necessary; what issues are being targeted and why; the parameters of the study; how results are going to be evaluated; and what standards are going to be used as a measuring aid;
- give details of monitoring systems to be installed and the contingency plans to be in place when 'trigger levels' are reached;

ii) statutory consultees are consulted in detail at the pre-application stage and then formally on receipt of the application;

iii) the authority encourages the applicant to create an early dialogue with the public, ideally before the application is submitted;

iv) use by the authority of specialist advisers to help determine the adequacy and carry out an evaluation of the ES once the application and ES have been submitted. Usually the authority seeks to get the applicant to pay for this service.

Four other authorities use the Institute of Environmental Assessment to determine the adequacy of the information provided in an ES.

1.7 **Pre-application Discussions/Scoping:** Nineteen of the authorities (35%) seek to enter into pre-application discussions to determine the scope of an ES where possible. About a third of these provide a checklist of the areas to be covered by the applicant in the preparation of the EA. Five authorities carry out an in-house scoping exercise independent of discussions with the developer, and one of these authorities carries out its own site survey to determine the issues likely to be raised by the application. In a further four cases the authorities rely on their own experience to determine the likely main issues at the pre-application stage.

1.8 **Use of Published Guidance:** Published guidance was identified as being of some help in determining the adequacy of ESs and evaluating the environmental information. The number of authorities using the various guidance available breaks down as follows:

-	The 'Blue Book'	22	41%
-	Circulars	10	19%
-	Kent CC Handbook	3	6%

The percentages do not add up to 100 because some authorities rely on the Regulations while others refer to more than one of the sources listed. The Kent County Council Handbook has apparently proved to be particularly useful for authorities within Kent. Also, it is acknowledged that the Regulations are not strictly guidance because they set out statutory requirements. Nevertheless local planning authorities use Schedule 3 as a checklist.

1.9 **Use of Statutory Consultees:** This is perhaps the area where there is most consistency between the various approaches adopted by authorities. A clear message emerging from the telephone survey was that authorities rely quite heavily on statutory consultees, and, to a lesser extent, on other consultees, to help determine the adequacy of, and evaluate, the information provided in an ES (33 [61%] of authorities stated that this was an important part of the process, although they also said that this would usually be the case for applications without an ES). The tasks of determining the adequacy of, and evaluating, the information could not easily be distinguished because they tend to be carried out simultaneously. Where they know about the project early enough, nine of the authorities specifically consult statutory consultees at the pre-application stage.

1.10 The reliance placed on statutory consultees for their opinion on the ES raised some concerns for some officers. Five authorities complained that statutory consultees are too parochial in their assessment of ESs, only looking at the areas within their remit and not paying enough attention to the wider issues raised by the application. This is not always helped by the practice of some authorities of only sending those parts of the ES relating to the area of expertise of the statutory consultee for comment. Other planning authorities felt that some of the statutory consultees failed to provide sufficient guidance on what are often highly technical areas which, although often governed by other environmental protection legislation, may have a considerable bearing on the final planning decision.

1.11 **Use of Other Consultees:** Many authorities found the advice provided by non-statutory consultees useful. Nineteen (35%) regularly use such bodies to give specialist advice and comments on an ES. The organisations most often mentioned were the county wildlife trusts, although national bodies such as the RSPB have also been consulted on occasions.

1.12 **Public Consultation:** The approach adopted to involve the public seemed to vary according to the particular application being considered. The more sensitive applications usually provoked greater public interest (and controversy) and in these cases, consultation over and above the statutory requirements was considered desirable by the planning authorities. Eight (15%) authorities encouraged the applicant to hold public meetings or an exhibition, although others were aware that the applicant had provided information to local residents, sometimes in the form of leaflets.

1.13 **Use of Other Specialists/Consultants To Evaluate ESs:** Eighteen (33%) of the authorities interviewed have used external and/or `in-house' specialist consultants to evaluate ESs in the past. Some authorities which had not used consultants said that they might do so in the future, although there was a general feeling of concern that resource constraints may prevent his happening in practice. Some authorities stated that they could not afford consultants in any circumstances.

1.14 The most common topics requiring input from consultants were noise and pollution. In particular, external consultants have been used to assess information on emissions from proposed incinerators. In addition, the two authorities involved in determining airport applications used outside consultants to help evaluate the noise implications.

1.15 **Evaluation Methods and Techniques:** The representations made by statutory and non-statutory consultees and, where used, the comments of specialist advisers, were seen as an important part of the evaluation process. Two authorities stated that they had used simple cost benefit analysis techniques to help with aspects of their evaluation, but otherwise no monetary evaluation method had been used.

1.16 Comparison with established standards (e.g. water quality standards) was the most common approach to non-monetary evaluation identified. Otherwise, one authority had used a computer model to validate the results in an ES and another used pair-wise comparison techniques for appraising road alignment options.

1.17 Although a few authorities are positive in their attitude towards employing techniques in the future, most are extremely doubtful about their usefulness. There appear to be two reasons for this. First, a general scepticism as to how environmental assets can be valued and whether an authority has the expertise and resources to carry this out. Second, planning recommendations are a matter for professional judgement and it is unlikely that there is a technique sophisticated enough to replace this.

1.18 **Presenting Information to Planning Committee:** A wide range of methods have been adopted by planning officers when presenting information to Committee members for decision. Generally, the Committee Reports tended to be longer than for a standard application because they summarised the main aspects of the ES. Some authorities try to present as much information as possible, including copies of all the representations, consultants' reports, the ES itself and, in two cases, the presentation of a seminar to members explaining the issues raised. One authority encourages site visits by the members.

1.19 Other authorities adopt a different stance, preferring to prepare brief summary reports of the environmental information, sometimes incorporating the non-technical summary from the ES. One authority attempts a form of cost benefit analysis to present information to members whilst another summarises the information in a matrix.

1.20 The general conclusion is that there is no one preferred method adopted by authorities and that the method chosen tends to depend on the political sensitivity of the application and the presentation techniques found to be most useful by the members of a particular authority.

1.21 **Resources:** Fifteen (28%) authorities said that at least one member of staff in the planning department had been on training courses covering EA, although there was some concern that short courses do not provide the detailed coverage that planning officers require. Ten (19%) authorities expressed particular concern about the lack of time available to assess an ES and eleven (20%) believed that they lacked the expertise to deal with ESs, hence the dependence placed on responses from consultees.

1.22 **Areas of Concern:** In addition to the points raised above (paragraphs **1.10, 1.13** and **1.21** for example), several authorities expressed concerns related to their ability to process ESs. These are summarised in **Table 1.4** for information.

Table 1.4 - Areas of Concern

Area of Concern	Number of Authorities
Lack of objectivity of ESs	21
Lack of guidance for interpreting technical information	5
Statutory consultees too insular	5
Impossible to standardise assessment of ESs	4
Difficulty in validating information	2
Lack of guidance on how to carry out scoping	2
Lack of environmental standards against which to judge the significance of environmental effects	2
Excessive size of ESs and the prohibitive cost for public	2
Lack of guidance on defining 'significance'	1

1.23 In addition, the use of consultants to prepare ESs prompted interesting comments from some authorities. Although responses suggested there is a tendency for authorities to give ESs prepared by consultants rather than by developers themselves more credence, there are still concerns about validating information. Some authorities felt under pressure to accept the findings of a consultant due to the time and money spent by the applicant in commissioning an ES. There was also some concern that the planning authority is not always in a position to understand the consultants' findings, especially where they deal with complex technical issues. A solution suggested by two authorities is for them to be involved in selecting and briefing the consultant, with the applicant footing the bill for the consultant's input.

APPENDIX 3 : SUMMARIES OF DECISION MAKING METHODS

(i) Cost-Benefit Analysis

(ii) Planning Balance Sheets

(iii) Goals-Achievement Matrix

(iv) The Integration of Environmental Assessment Into Planning

(v) Pair-Wise Comparison

(vi) Decision Analysis

(vii) Safe Minimum Standards

COST-BENEFIT ANALYSIS

Recent refs - Jacobs,M. (1991) 'Making environmental decisions (1): The limits of cost-benefit analysis' (chp16 pp195 - 203) in *The Green Economy*, and Dixon,J Sherman,P (1991) *Economics Of Protected Areas - A New Look At Benefits And Costs.*

Introduction

Cost-benefit analysis (CBA) attempts to assess the desirability of alternative projects/policies by measuring and then comparing their costs and benefits. Both private and social (including environmental) costs and benefits are considered. CBA starts from a simple premise, namely that an investment project should only be undertaken if all its benefits outweigh all its costs. If a project has several alternative forms, or there are competing projects, providing that the benefits are greater than the costs, the one for which the difference between benefits and costs is greatest should be chosen. As costs and benefits can only be added and subtracted if expressed in the same units, CBA attempts to place monetary values on them as money is a convenient 'measuring rod' of value. Since the costs and benefits of a project will occur over a period, future costs and benefits are subjected to a discount rate which converts them into their 'present value'

CBA as a method has a relatively long history. Although originating in France during the mid 19th century in the writings of Dupuit, it really came into prominence in the USA in the 1930s when it was applied to their vast investments in water resource projects. Perhaps the most notorious of the CBAs carried out in the UK, was that undertaken by the Roskill Commission in the early 1970s as part of their search for a third London airport. Their attempts to place monetary values on social and environmental variables met with much criticism and controversy that still continues today.

Methodology

After the project and its alternatives have been clearly defined (preferably to include a 'do-nothing' option), the respective costs and benefits of the various alternatives must be identified, listed, and then given a monetary value. Where no market exists (as is the case with many environmental variables), substitute or 'shadow' prices must be assigned to allow financial appraisal. Techniques employed to do this include hedonic pricing (which, for example, looks at the extent to which property values are affected by environmental variables), and contingent valuation (which basically asks people what they are willing to pay for a benefit, and/or what they are willing to receive by way of compensation to tolerate a loss). The travel cost method infers the monetary value of recreational areas, for example, by investigating how far people are prepared to travel and how much they are prepared to pay to visit such an area.

The next stage is the discounting of these monetary values to give a present value using an appropriate discount rate (the selection of which can itself be open to much debate). The results of this process are presented as a rank order of alternative schemes in terms of, for example, their benefit / cost ratio, although this may not reflect absolute values which may considerably differ. The intangible items (often including environmental factors) should preferably be considered separately, with a full explanation of all assumptions made at the various stages of the analysis.

Applicability to the Planning Decision

CBA is a method that can be used in either the decisions that involve the appraisal of alternative options or those determining whether or not one project should be implemented. Whilst useful in that it brings an economic dimension to environmental considerations, the accuracy of the monetary valuation techniques associated with CBA is questionable. In theory, CBA would be most useful where the factors under consideration could be 'sensibly' monetised with relative ease. The application of CBA and its associated monetary valuation to 'emotive' variables' such as areas of high landscape value or endangered species would be open to considerable criticism.

Practical Implications

Whilst some EAs may incorporate some degree of cost-benefit analysis, it is more likely that the significant environmental considerations identified in an ES would be able to contribute to the list of variables to be priced and then compared in a CBA. However, the monetary valuation is likely to require considerable skills and experience above and beyond those commonly needed for the decision making process. A considerable amount of additional information is required to carry out a CBA, such as up to date property values, traffic flows, and social surveying of peoples preferences that is both reliable and appropriate to the case under consideration. Aspects of this process are both time consuming and technically difficult, and whilst CBAs can be carried out by 'experts' external to the planning system, the assumptions and values they incorporate into the valuations may not necessarily correspond with the aims and objectives of the planning authorities themselves. Moreover, although a CBA could be carried out in parallel with the preparation of an ES, completing a CBA solely in the standard sixteen week period between receipt of an ES and the need for a planning decision would be impractical.

PLANNING BALANCE SHEETS

Refs. - Lichfield , N. (1970) ' Evaluation methodology of urban and regional plans : a review' in *Regional Studies* vol. 4 pp151 - 165. Ratcliffe, J. (1981) ' Evaluation' (chp 14 pp249 - 264) in *An Introduction to Town and Country Planning,* 2nd edition.

Introduction

Planning Balance Sheets (PBS) is a method closely associated with cost-benefit analysis (CBA). It was first put forward by Professor Nathaniel Lichfield (University College London) in 1956 and attempts to provide a practical way in which CBA can be applied in town planning and development. Although primarily devised to evaluate alternative plans, it is possible that the method could be applied to evaluating alternatives at the project level.

Unlike many typical CBA studies, the PBS cannot and does not aim to provide a conclusion in terms of rate of return or net profit measured by monetary values. Instead its objectives are to expose the implications of each set of proposals to the whole community, and also indicate how the alternatives might be improved or amalgamated to provide a better result The purpose of the approach is the selection of a plan (or in this case project), which on the information available is likely to best serve the total interests of the community (Lichfield 1970).

Methodology

The PBS sets out by identifying two broad categories of individuals/groups within the community, these are the producers responsible for introducing and operating the particular project, and the consumers who will be the recipients of its effects. Having determined the respective membership of these two sectors, the costs and benefits that accrue to them are compared and valued. Those that are capable of being assessed in money terms are entered into the *balance sheet of social accounts,* with the costs and benefits divided into both annual and capital. In the same way, costs and benefits that are considered to be intangible and defy monetary quantification are entered as *I* if they are capital items and *i* if they are annual. Similarly, those that are measurable, but because of lack of time, expense or lack of data are not measured, are entered as *M* or *m* respectively. Subscripts are used to distinguish between the different M variables, for example M_1 may refer to developed land, and M_2 to land that is undeveloped. If an increase over the existing 'do-nothing' situation is envisaged then a positive (+) sign is employed; conversely if a decrease is predicted, a negative (-) sign is used. Some attempt is made to estimate the magnitude of these unquantified variables, for example, M^{++} is used to denote an amount greater than M^{+}. In this way, each alternative plan is compared against the 'do-nothing' situation.

The next stage is the 'reduction stage', which aims to eliminate not only double counting (a common problem in CBA), but also transfer payments and common items of cost or benefit. The reduction sheet subtracts all items entered more than once, which are marked with an *E*, and all items of equal worth in each alternative. In the table overleaf for example (adapted from a Peterborough case study by Lichfield, 1969), M_4 is removed from the PBS at the reduction stage as it has a value of M in all options. The remaining items are summed algebraically, and reduced to a common form, either annual or capital. The degree of change between alternative plans can be gauged by comparing the various elements one against another, and arriving at either net advantage to one plan, or, where several alternatives exist, an order of preference can be established. In some studies (see for example N.Lichfield 'Cost-benefit analysis in urban expansion : a case study - Peterborough, *Regional Studies*, vol.3, 1969), this method is further developed by attempting to rank or weight the various elements such as the relief of congestion or the displacement of population. Whilst this is acknowledged as involving a considerable amount of value judgement and political 'interference', it is believed to lead to far more realistic estimates if properly conducted.

ITEM	PROJECT 1	OPTIONS 2	3	4
LAND				
1. Developed	M_1++	M_1+	M_1	M_1
2. Undeveloped (agricultural)				
a, Good quality	£374,500	-	-	-
b, Average quality	£756,000	£898,000	£325,500	£345,000
c, Below average quality	-	-	£373,500	£373,000
d, Potential for development	£186,000	£223,000	£153,500	£143,000
e, Clay workings	£12,000	£16,000	£31,500	£29,500
f, Development value	M_2++++	M_2+++	M_2+	M_2
g, Open space	M_3	M_3	M_3	M_3+
MAIN SERVICES				
1. Storm water	M_4	M_4	M_4	M_4
2. Gas	M_5	M_5	M_5+	M_5++
3. Electricity	M_6	M_6	M_6	M_6+
REDUCTION PROCESS				
	£1,328,500	£1,137,000	£884,000	£890,000
	M_1++	M_1+	M_1	M_1
	M_2++++	M_2+++	M_2+	M_2
	M_3	M_3	M_3	M_3+
	M_5	M_5	M_5+	M_5++
	M_6	M_6	M_6	M_6+

Applicability to the Planning Decision

PBS is quite flexible in that it does not attempt to price all variables, so could perhaps be used for evaluating projects that have significant conservation implications, for example. It is also flexible in that weighting can be added but is not always assumed. It is capable of either comparing one project against the 'do-nothing' alternative, or of comparing a number of alternatives against each other, such as alternative routes for a road.

The PBS was devised prior to the introduction of EA in the UK, so it is not designed specifically to relate to the information that is usually provided through EA procedures. It evaluates the different project alternatives from the point of view of various community sectors, but as it was developed prior to EA, it does not explicitly make reference to the views of statutory consultees such as English Nature. In addition, although the PBS could be compiled internally by the decision makers, it does not necessarily highlight which options conform closest to the stated or implied objectives of the planning authorities.

Practical Implications

To complete a PBS, information is required on the different 'community sectors'. Some of this may be contained in the 'social impact' section of an ES, but additional census type information on population sizes and occupations, for example, might be required in order to categorise these community sectors. Unlike the CBA, however, this method does not need information such as social surveys as it does not attempt to judge the preferences of the different community sectors regarding different development options or try to select an alternative on their behalf.

Although variables can remain unquantified, some indication of their magnitude is needed, for example whilst the creation of recreational space does not have to be stated in acres, it is necessary to know whether one option will result in a greater increase (M^{++}) than another (M^{+}). Additional financial and mathematical information, skills and experience are required to apply monetary valuations, and although the process of summing costs and benefits algebraically, and then 'reducing them to a common form' does not require any great mathematical knowledge, it may require a more mathematical way of thinking and presentation of information than commonly applied in planning.. Additional information and skills may also be necessary to apply any weighting, and also perhaps to determine whether the impacts of the project are an improvement or not when compared to the original situation. Although this is only necessary where considered 'appropriate', the determining of this 'appropriateness' may in itself require certain information and skills.

One advantage of this approach that perhaps makes it more realistic than some of the other methods, is that it takes account of cases where it is not possible to apply monetary values (*I* and *i* values), and those that are not measured due to lack of time, expense or lack of data (*M* and *m* values). Despite this, whether such a balance sheet could be compiled with a fair degree of accuracy, during the standard sixteen week period between submission of Environmental Statement and the planning decision for example, is debatable. Another option is that such a balance sheet is actually compiled as part of the EA process leading to preparation of the ES.

GOALS-ACHIEVEMENT MATRIX

Refs. - Hill, M. (1968) 'A goals-achievement matrix for evaluating alternative plans', *Journal of the American Institute of Planners*, 34, 19-29. Hill, M. (1973) *Planning for Multiple Objectives An Approach To The Evaluation Of Transportation Plans.* Ratcliffe, J. (1981) 'Evaluation' (chp 14 pp249 - 264) in *An Introduction to Town and Country Planning*, 2nd edition.

Introduction

The Goals-Achievement Matrix (GAM) was developed by Morris Hill in a doctoral dissertation presented in America in 1966. Although Hill devised this method to establish the extent to which a plan meets the original objectives set by politicians, it could be adapted for decision making at the project level - for example to compare project implications against plan objectives. Hill developed his method after criticisms that both cost-benefit analysis and the planning balance sheet tend to classify all the costs and benefits of alternative strategies with no regard to the politics of a particular situation. He argues that talking 'of maximising net benefits in the abstract is meaningless', and that if for example, the community in question sets no value on the retention of historic buildings, it is not legitimate to consider the elimination of a building of historic value as a cost for that community. Hill suggests that the goals and objectives of a scheme should be made explicit and that the alternative strategies be measured in terms of the extent to which they achieve them.

The constituent elements of the plan (or project), such as housing, employment or open space, are weighted before a matrix is constructed to indicate the political preference or priority attached to them. The various agencies or groups that will be affected by the proposal are also weighted to reflect their preferences and the political pressure they exert. In this way a matrix showing the relative performance of difference options with respect to the defined objectives is constructed. Following this, social accounts are drawn up in a similar way to those of the planning balance sheet .

Methodology

The term 'goal' is used generically and is defined as 'an end to which a planned course of action is directed'. Goals may involve gaining something not already obtained or removing something undesirable that is already there, and they can be categorised at three levels of detail. *Ideals* allow for indefinite progression in their direction such as increased economic welfare or improved environmental health. On a more detailed scale,*Objectives* denote attainable goals that often do not have intrinsic value themselves but have instrumental value in that they can lead to another valued goal. An example is the reduction of air pollution (an objective) that can contribute to improving environmental health (an ideal). Finally, *policies* are the most clearly defined of goals, and relate to the specification in precise detail of ways and means for the attainment of planned objectives, such as the separation of rights of way for pedestrians and vehicles. Hill argues that the appropriate level of detail for the purposes of the matrix, involves defining goals at the level of objectives. The costs and benefits of the plan or project should then be entered into the matrix under the goals they relate to, and the degree of achievement of the various objectives measured. The greater the benefits of introducing clean air technology for example, the greater the degree of achievement of improved environmental health standards.

Evaluation of alternative courses of action requires the determination for each alternative of whether or not the benefits outweigh the costs, so for each alternative, a matrix is constructed. Within each matrix, a 'cost-benefit account' is prepared for each goal, with some tangible costs and benefits that can be expressed in monetary terms, other tangible costs and benefits which can not be expressed in monetary terms but can be defined quantitatively, and intangible costs and benefits that can only be defined qualitatively.

The decision maker then has to weigh the alternative courses of action against each other, with the weighting reflecting the community's valuation of each of the objectives and the incidence of benefits and costs. It is therefore necessary to identify those sectors of the population who are likely to be affected by the consequences of a course of action since the consequences are unlikely to affect all sectors. Allowance should be made for uncertainty (for example, through the use of conservative estimates and the requirement of safety margins), and for the time dimension whereby costs and benefits occurring in different time periods are not of equal weight.

An example of a goals-achievement matrix (adapted from Hill 1968) is shown below :

Goal Weight	I Two			II Three			III One		
	Weight	Costs	Benefits	Weight	Costs	Benefits	Weight	Costs	Benefits
group a	1	A	D	5	E	-	1	Q	R
group b	3	H	-	4	-	R	2	S	T
group c	1	L	J	3	-	S	3	V	W
group d	2	-	-	2	T	-	4	-	-
group e	1	-	K	1	-	U	5	-	-
		----- t	----- t		----- t	----- t		----- t	----- t

In the above table, *I*, *II*, and *III* refer to the goals (as defined by politicians), such as the reduction of air pollution or the preservation of historic buildings. Each goal is weighted (*one, two and three* in the table), and various groups of people (groups *a,b,c,d* and *e*) are identified as affected by the course of action. A relative weight is then determined for each group, either for each goal individually or all goals together (*1,2,3,4* and *5*). The letters *A, B, C*.....are the costs and benefits that may be defined in monetary or non-monetary units or in qualitative terms, for example *A* could be £10, 5 units of air pollution or simply a reduction in landscape quality, and there is no indication of whether the cost of *A* is greater than the cost of *B*. A dash (-) in a cell implies that no cost or benefit that is related to that objective would accrue to that party if that plan or project was implemented. For certain of the goals *t* indicates that summation of the costs and benefits is both meaningful and useful, and that the total costs and benefits with respect to that goal can then be compared. This will be the case when all the costs and benefits are expressed in quantitative units, but will not be the case for intangible costs and benefits relating to the achievement of qualitatively defined objectives. As it is unlikely that all the costs and benefits of all the goals can be expressed in the same units, a final cost-benefit summation is possible but will only occur in rare cases, perhaps where only one or two goals are valued by the community.

Hill tackles the issue of comparing objectives that are expressed in qualitative terms with those measured quantitatively by arguing that any value that can be quantified, can also be expressed in qualitative terms, and that scales of measurement can also be used for comparative purposes. For example, an ordinal scale can be used whereby a value of *+1* is allocated where goals-achievement is enhanced, *-1* if goals-achievement is decreased, and *0* if there is no effect on goals-achievement. Weights can then be assigned both for the goals and the population sectors as in the matrix example given here, and a total figure derived for each goal from combining the +1, -1 and 0 scores for each plan / project.

Applicability to the Planning Decision

The Goals-Achievement Matrix is another method that was devised prior to the formal adoption of EA in the UK, so does not tie in directly with the environmental information provided in an Environmental Statement. Whilst much of the environmental information necessary to prepare a GAM should theoretically be supplied through the EA procedure, in reality this is unlikely to be the case, and additional information may have to be requested.

Hill himself believes that, as in the case of cost-benefit analysis and the planning balance sheet, the goals-achievement matrix cannot determine whether a project should be executed or not. It is designed primarily for the comparison and ranking of alternative projects rather than for testing their absolute desirability. As all the planning applications accompanied by EAs require a 'yes-no' decision, the goals-achievement matrix has, like the other methods just cited, only limited applicability

Whilst the objectives and value judgements of the decision makers are explicitly stated with this method and also have the potential to take account of interests other than those that are strictly human-orientated, difficulties may arise because the goals may be conflicting and also ultimately unattainable. In addition, there is no obvious way of incorporating the goals of the statutory consultees for EA, who may have a different emphasis from those of the planning authorities.

Practical Implications

One advantage of this method is that, like the planning balance sheet, it is not necessary to define all values in monetary terms. However, even deciding if the project moves towards or away from a defined goal may not be straightforward and could therefore have resource implications, as would allowances for uncertainty and the time dimension.

Although census-type information can be used to define the community groups, perhaps the factor that would most limit the practical use of this method in planning decisions is the need for the weighting of both community preferences and political aims, which does not normally occur and would require considerable time and expertise that may not be available. It is, for example, likely that specific social surveys taking into account both the project implications and the politically defined goals in consideration would have to be carried out.

Hill himself admits that 'this method calls for an extremely complex, time consuming, and expensive task' and also that 'it may be contended that the product of the goals-achievement matrix is exceedingly complex and thus difficult for the decision maker to digest' (Hill, 1968). A goals-achievement matrix could not realistically be completed in the sixteen week period between receipt of an ES and the need for a planning decision, and as much of the environmental information will not be available in an appropriate form until the ES is completed, little work on the matrix could be carried out prior to the submission of this document. Therefore, whilst perhaps being conceptually pleasing, the goals-achievement matrix would appear in practice to be of limited use.

171733 E

THE INTEGRATION OF ENVIRONMENTAL ASSESSMENT INTO PLANNING

Refs - Lichfield,N. (1992) 'The integration of environmental assessment into development planning : part 1, some principles' in *Project Appraisal* vol. 7 no. 2 pp58-66. Lichfield,N (1992) 'The integration of environmental assessment into development planning : part 2, a case study' in *Project Appraisal* vol 7 no 3 pp175-185.

Introduction

The Integration of Environmental Assessment Into Planning is a method devised in recent years by Nathaniel Lichfield (Professor Emeritus of the Economics of Environmental Planning at the University of London) in response to what he believes is the inadequate integration of environmental assessment into planning to date. Lichfield argues that the 'disciplines' of planning and environmental assessment have much to offer each other, and that this has yet to be effectively capitalised upon. EA, he suggests, is capable of bringing to planning an emphasis on systematic analysis, that uses the best available techniques and sources of information. In addition, EA presents information in a form that provides a focus for public scrutiny of a project, and enables the importance of the predicted effects and the scope for modifying/mitigating them, to be properly evaluated by the planning authority before a decision is given. Planning is believed to provide a focus for EA, by emphasising the role that EA is designed to fulfil. A need is identified for the integration of environmental scientists who must themselves become 'planning orientated', as opposed to simply making distinct and expert reports on environmental problems.

Lichfield believes that the DoE has taken a strong lead in incorporating planning into EA (ref. Policy Appraisal and the Environment, DoE 1991), so his technique is one that attempts the converse, incorporating EA into planning.

Methodology

The first stage is the establishment of both the need for the project under consideration, and the justification for the chosen site. The project is then described, including reference to potential effects both before and after mitigation. Two options are considered: the proposed development with mitigation (option A); and the same site if planning permission is refused, and the development does not go ahead but the site is subject to any mitigating or enhancement measures which may be appropriate as a result of its former use (option B). The development proposal is split into three phases: construction to implement the development; further physical development after completion (such as parking and water features); and operation and management post completion. The potential environmental effects of each phase are then identified and assessed, including, for example, those on flora and fauna, air and water. Provided that an EA is required by law for the project under consideration, the stages of assessing environmental effects would be present in an ES that satisfies the statutory requirements. This would not necessarily be the case for details concerning the site without the project, but with any appropriate environmental mitigation or enhancement.

The distinction is made between environmental *effects* (physical and natural changes) and the actual *impacts* on human beings, which are the critical aspects when making planning decisions. Since the impacts on human beings are not uniform, Lichfield divides up human populations into community sectors such as landowners, operators and consumers. Each environmental effect of the three development phases is then divided up to determine its impact on various community sectors. For example, a change in air quality is split into its impacts on consumers and its impacts on operators. These community sectors are further subdivided into 'onsite' sectors (such as the National Rivers Authority) and 'offsite' sectors (such as the Dept. of Transport). Once the environmental impacts of the three phases on all the subdivisions of the community sectors have been described, the preference of these sectors for the site either with development (option A) or without development (option B) is 'postulated'. In the case study example provided by Lichfield (a proposal by British Airways for creation of a Headquarters and Business Centre at Heathrow airport), these preferences are derived from a combination of common sense and interpretation of a social survey commissioned by British Airways. Because each community sector is delineated by reference to a particular environmental aspect, any one individual or family can be present in a number of sectors.

This method requires the identification of the community sectors affected by each environmental effect, but as planning assessment requires that the effects are attached to a particular sector so that decisions can be made on behalf of a community, this process has to be reversed and the impacts grouped by sector. For example, the impacts of both option A (development) and option B (no development) on British Airways as an operator, are listed, and then a preference for either option A or B inferred. This is carried out for each of the three stages of the development, with any existing uncertainty being noted. From a comparison of all these As and Bs, conclusions are drawn as to both the overall environmental impacts and those impacts on human beings, for all the three phases, and then the final decision made as to whether the net benefits are to be found in implementing or not implementing the project.

Applicability to the Planning Decision

This method has the advantage of being specifically devised to incorporate EA, rather than being created prior to EA, and then modified later to take account of EA planning requirements. Whilst designed to aid the decision as to whether or not a particular project should be implemented, there is no reason why several alternative projects or locations could not be compared, (by the addition of *C*s and *D*s for example).

Like EA, Lichfield's method does break down the development process into different phases and areas of impact which present a clearer and more logical picture and in this way aid decision making. In addition to this, the views of different community sectors are explicitly recognised, which could in turn increase public accountability. For example, if residents close to a site where planning consent has just been granted, complain that their views have not been considered, the use of Lichfield's method could in some cases reveal that their preference for no development (option B) was recognised, but that there were many more community sectors that preferred the development to go ahead (option A), and because of this the development was granted permission.

One problem with this method however is that environmental effects are evaluated solely by reference to their impact on human community sectors, which may not comply with the overall objectives of the planning authorities. When Lichfield's method is applied to a particular site for example, the selection of an option for development of a retail centre (rather than the 'do-nothing' option) may be chosen on behalf of the community sectors, because it provides a convenient location for a facility they desire. This may occur despite the resultant loss of several endangered species and an ancient woodland, and the fact that there are several other alternative sites only slightly further away from the population centre, where no such loss would be incurred.

Decision-making based on this method would be easier where the results clearly favour one of the options (in other words where there was a clear predominance of either As or Bs), as is the case in the Heathrow example cited by Lichfield. If this is not the case (and there is largely a mixture of As and Bs), it may be necessary to introduce some element of weighting. Although this is acknowledged by Lichfield, who highlights the same situation where the preferences of those close to the development differ from those further away, no advice is given on how such weighting could be carried out if the situation was not as 'clear-cut' as in the case study provided.

Practical Implications

One drawback with this method is that it requires social survey data that is both reliable and appropriate to the case in hand, and the carrying out of such surveys can be difficult, time consuming and costly. Existing social surveys can be utilised, but they need to be targeted to that particular situation, and if carried out externally (by the developer for example) they should be examined for possible bias first. Even if the surveys are appropriate, they still need to be interpreted in considerable depth to allow preferences to be inferred, which again can take a considerable amount of time and require certain skills and experience. Whether such surveys could be carried out, interpreted and incorporated in the usual time available between receiving an ES and coming to a decision over the granting of planning consent is debatable.

PAIR-WISE COMPARISON

Refs - Lichfield,N. (1970) 'Evaluation methodology of urban and regional plans: a review ' in *Regional Studies*, vol 4, pp 151-165. Land Use Consultants (1992) 'Examination of alternative locations for landfill waste disposal in south Norfolk'. Kass,G.S. Walker,C.E. (1992) 'Pairwise analysis : a tool for comparative EA', Paper from the International Conference on Environmental Assessment 1992, sponsored by the Institute of Environmental Engineering, University of Nottingham.

Introduction

The method of pair-wise comparison was first developed by the US Army Corps of Engineers. It was used after the second world war, as part of the process of evaluatng alternative dam sites in the Tennessee Valley. It is a procedure that attempts to progressively rank the desirability of alternative options, by directly comparing them against each other. By taking two options and then comparing their strengths and weaknesses with respect to particular criteria, each option is compared to all of the other options, to see which option, or set of options, performs best overall. The preferred options are then taken forward for further evaluation, whilst the less favoured options are put to one side or rejected altogether. In practice, the ideal choice may not materialise, especially if there is uncertainty about the degree of confidence with which individual judgements have been made. In such cases it is necessary to study further those factors that seem to discriminate between the most favoured group of options, and then to repeat the comparison when more detailed information is available.

Methodology

The first stage involves defining all the options under consideration, such as the different projects proposed for one site or all the potential locations for a certain development. The criteria considered to be of importance when evaluating and comparing the different options then have to be identified. These criteria may be limited to environmental factors, or may include technical and economic factors. The latter is particularly likely if pair-wise is incorporated by a developer, in feasibility studies that look at project alternatives for example. Social factors may also be included, and are perhaps of particular importance to planning authorities charged with making decisions on behalf of sectors of the population.

These different criteria could be considered using separate and distinct pair-wise processes, to select options most preferable on environmental, economic, social or technical grounds. Alternatively, attempts could be made to incorporate all these factors into one 'master' list for comparison. Some of these may be costed, such as the loss of agricultural land, whilst for others such as damage to landscape quality, this is unlikely.

For each criterion selected, the impacts of the different options are then assessed, ideally incorporating not only the severity of the likely impact but also whether or not it could be substantially mitigated. These impacts can then be ranked. To take impacts on water quality for example, in a situation where there are ten possible sites for waste disposal, the site where the impacts on water quality are least will be ranked *1* (most preferred), and where impacts on water quality will be greatest, *10* (least preferred). This is repeated for all the criteria, so that if there are 15 criteria under consideration, each option will have a series of 15 values that range between 1 and 10, (although the criteria may have to be subdivided, such as 'ecology' into flora and fauna, and 'social' into job creation and provision of new services). Kass and Walker (1992) in their case study however, choose not to rank the impacts in an attempt to remain more objective. Instead they allocate a value of 'low', 'medium' or 'high' to each impact, based wherever possible on quantitative criteria.

The standard procedure then involves selecting one of the criteria that is believed to be most important, and then comparing the different options against each other, with respect to this criteria. Out of the first two options that are compared with respect to ecology for example, the option that would involve building on a nature reserve would either be discarded, or ranked below the option that did not. The preferred of these two options is then compared with another alternative. This process continues until all the options have been compared in terms of their ecological impacts, with those having significant ecological impacts perhaps being put to one side or discarded. Another important criterion such as landscape quality is then selected and the process repeated with the remaining potentially 'suitable' options. The idea is to compare alternatives across the board on ecological grounds and then on landscape grounds, rather than directly weighing up the ecological impacts of one option against the landscape impacts of another and then ranking the two options on the basis of this result.

The different criteria may themselves have to be weighted however to determine the order in which the impacts are compared, For example whether ecology is in this case more important than landscape, and therefore ought to be used as a first basis for comparison. Otherwise an option may be dismissed for having a poor ranking, when the criteria under consideration is actually relatively insignificant. In order to establish the sequence of this list, factors that should be taken into consideration include whether the impacts are likely to be of local or regional importance, short term or long term, and temporary or permanent.

Kass and Walker (1992), avoid this need for weighting, again as part of their desire for objectivity. In their example, the number of occasions where a particular option scores a higher level of significance in each pairwise comparison is noted. For example, in a pair-wise comparison between option A and B over ten criteria, option A may give rise to higher significance three times, and option B four times, with the remaining criteria showing no difference in significance. For each option, the total occurrences of these higher levels of significance are calculated across all pair-wise comparisons to give a 'preference rating'. The highest rating represents the option with the highest degree of severity of impacts, and indicates the least preferred option. In this way, the options can be ranked in order of preference.

Applicability to the Planning Decision

Pair-wise comparison is ideally suited to comparing different options, such as alternative routes for a bypass, or alternative sites for waste disposal. It can in theory be used to reach a 'yes or no' decision about whether or not a single option should be adopted, if the proposal is compared against the 'do-nothing' alternative. One advantage of this approach is that it avoids the need to make absolute judgements, which can be difficult where qualitative assessment is involved.

A pair-wise comparison may not however reveal the perfect option (such as one that reaches all the minimum accepted standards), as it will only rank the alternatives to give the one which is most preferred. In some cases, pair-wise can be used simply to reduce the range of options under consideration, rather than being 'trusted' to provide the single most desired alternative. This smaller range of options could then be subject to further evaluation if desired, by for example taking the 'Lichfield' approach and predicting the preferences of different sectors of the affected community for each of the shortlisted options.

Practical Implications

Although pair-wise comparison was devised prior to EA, it is a method that can incorporate the information provided in an ES quite effectively. It does however require detailed information on all the alternatives under consideration, including the 'do-nothing'. Although alternative options will ideally be mentioned in an ES, the emphasis of the Statement will inevitably focus on the option preferred by the developer; it is therefore likely that additional information will be required.

In addition, if the approach is adopted whereby both the severity of the impacts and the criteria must be ranked, considerable knowledge and experience may be needed in order to do this. This is especially important if subjectivity is to be minimised. Although the principles of pair-wise comparison are relatively simple, the collection and analysis of information on all the alternatives is likely to be time consuming. A pair-wise comparison could therefore only realistically be carried out between receipt of an ES and the need for a planning decision if, unlike at present, planning policy required that the environmental (and perhaps social) impacts of all the genuine project alternatives were considered in depth as part of the EA process.

DECISION ANALYSIS

Ref. The New Rational Manager, by C. H. Kepner and B.B. Tregoe, 1981

Introduction

Decision Analysis is a systematic procedure for deciding between various options. Although existing in different forms, this version was developed in North America by Kepner and Tregoe primarily to assist management decisions. However, LUC has used the method in a slightly modified form to evaluate possible flood alleviation channel alignments.

Methodology

The decision making process is broken down into a series of steps, as outlined below.

Step 1: The Decision Statement

The decision statement is a concise description of the decision that needs to be made. Its wording is important because it provides the focus for the analysis and indicates the level of the decision to be made. For example, it might be: to decide whether or not to grant planning permission for a project x at site y.

Step 2: The Objectives for the Decision

Objectives are clear statements of what the decision should achieve. They must be worded in a way that allows the alternative courses of action to be measured. There are two types of objectives. *Must* objectives are those that are mandatory and also measurable in absolute terms; for example, must accord with development plan policies. *Want* objectives are those that are not measurable in absolute terms; for example, must minimise environmental costs. Must objectives are attached more importance than want objectives - must objectives specify which options should be considered and want objectives form the basis of further evaluation.

Step 3: Weighting the Objectives

Once the want objectives have been agreed, each one is weighted according to its relative importance. The most important objective is given a weight of 10. All other objectives are then weighted in comparison with the first, from 10 (equally important) down to possible 1 (not very important). No attempt is made to rank the objectives. The purpose of the 10-1 weighting scale is to make visible the relationships between these objectives. Only *want* objectives need to be weighted.

Step 4: Identifying Options

At this stage a set of possible options may already have been identified. If not, options must be generated. Ideally, this should be a creative process, involving as many different disciplines as possible. For example, possible road alignments could be identified in consultation with relevant specialists.

Step 5: Evaluating Options

Options are evaluated against the objectives. First the options are checked against the *must* objectives. Any option that does not meet a *must* objective can be eliminated immediately. The remaining options are then evaluated against the *want* objectives.

Evaluation and scoring against the *want* objectives is based on comparison of the available options. The option which best meets an objective scores the highest; the others are scored relative to that. This approach is important when the decision is to choose the best available option. A slightly different approach is used if looking for an "ideal" or assessing if any one option is "good enough". Once the options have been scored for each objective, weighted scores and a grand total can be calculated. These scores give an indication of how well each option performs against the whole range of objectives. High total scores will indicate which options are tentatively acceptable.

Step 6: Adverse Consequences Analysis

The group of high scoring, tentatively acceptable options must be examined more closely to discover any weaknesses. At this stage it is necessary to be very ruthless in examining each option and ask the following sort of questions:

- how was the option's score made up? Did the option score very badly in relation to one group of objectives?

- what internal organisational factors or changes could affect acceptance or implementation of the option?

- what external changes could harm the successful implementation of the option?

Each potential weak point is assessed in terms of the probability of the adverse situation occurring (high, medium or low). After this analysis it is necessary to investigate means of avoiding or mitigating potential problems, review the combination of components that make up the option, or eliminate the option entirely (possibly in favour of a lower scoring but less risky option).

Applicability to the Planning Decision

In theory, the method of Decision Analysis is applicable to planning decision-making. It is better suited, however, to the 'developer's decision' between various options. For a planning authority to use the method effectively, it would be necessary for the authority to define its objectives clearly and concisely, as a framework within which to make the decision. It would be possible however to translate the policies of the development plan into objectives against which to judge alternative options. *Must* objectives could be derived from policies seeking to protect the most valued areas (e.g. must avoid SSSIs), and *want* objectives could be derived from other relevant planning policies.

Once the objectives have been agreed, Decision Analysis could help to decide:

· whether or not to grant planning permission, where a refusal would be considered as the 'do-nothing' option;

· between different development options;

· whether or not to attach conditions to the preferred option (i.e. where mitigation is required if the option scores badly against one or more objectives).

Practical Implications

In practice, the method of Decision Analysis would be difficult to use in reaching a planning decision. This is because:

(i) it would be necessary to ensure that the planning authority's objectives are fully defined and agreed before the decision making process can begin;

(ii) the weighting of objectives could be difficult, politically contentious and cause delay;

(iii) it could require additional data for alternative options, in order to assess the extent to which each option meets the objectives (ie. to score them).

From our experience in using Decision Analysis in identifying a preferred flood alleviation channel alignment, it was the numerical nature of the weighting and scoring process that caused most difficulties for those involved in making the decision. People generally questioned the numbers that were derived and found the advantages and disadvantages of each option difficult to visualise.

In order to overcome these problems we modified the way in which the information was presented by using a colour-coded matrix, with environmental subject areas on the x axis and the different alignments on the y axis. The colour of the dots in the matrix denoted the importance of the objective (similar to weighting but measured as high, medium, low) and the size of the dot reflected the extent to which the objectives for each subject area (similar to scoring).

The principles underlying this approach remained faithful to the method of Decision Analysis and it provided a very useful way of highlighting the relative advantages and disadvantages of different flood alleviation channel options.

SAFE MINIMUM STANDARDS

Refs. S.V. Ciriacy-Wantrup, `Resource Conservation and Economics', 1986; J.A. Dixon and Paul B. Sherman, Economics of Protected Areas', Earthscan Publications, date?; M. Jacobs, `The Green Economy', Pluto Press 1991, R. Bishop, `Endangered Species and Uncertainty: the Economics of a Safe Minimum Standard, American Journal of Economics, 61(5) 1978'. Barker, A., Bowers, J., Hopkinson, P. and Lyall, K. (1993) Environmental Standards : Issues for the Countryside Commission, Environmental Policy Unit, University of Leeds, Report No. 1, Draft Final Report.

Introduction

The concept of `Safe Minimum Standards' was introduced in 1952 by S.V. Ciriacy-Wantrup. It is based on the proposition that critical resources (he cites soil, water, plants and animals) should be protected at some minimum level as a buffer against both unexpected environmental behaviour and the consequent risk of irreversible change. He refers to a critical zone, beyond which irreversible depletion of the resources will take place, thereby limiting the potential development of society.

More recently the concept has been expanded to embrace a wider range of environmental resources and has gained popular support under different names, eg. environmental constraints/standards/thresholds and sustainability criteria.

In a recent report for the Countryside Commission, the University of Leeds Environmental Policy Unit identifies two principal types of environmental standard:

(i) Discharge standards, that are applied to discharges of pollutants to the receiving media of air, water and land. The DOE's statutory water quality objectives provide a contemporary example.

(ii) Stock standards, which are locationally specific (eg. SSSI's, scheduled Ancient Monuments).

The application of such standards is seen as a means of making the concept of sustainability operational (Jacobs, 1991).

Methodology and Application to Planning

The concept of Safe Minimum Standards is not, in itself, a methodology for making planning decisions. It can, however, provide a very useful `first sieve' in the consideration of development proposals, and can be used in conjunction with other methods to reach a final decision.

Practical Implications

The use by local planning authorities of Safe Minimum Standards, or similar concepts under a different name, is severely limited by the fact that, in relation to many indicators of environmental quality, minimum standards have yet to be set. This is because our knowledge of the environment is far from perfect. Considerable work would be required to define Safe Minimum Standards in relation to a development proposal.

In response to this problem, the `precautionary principle' is widely advocated. This is defined by the DOE as `a decision to take avoiding action based on the possibility of significant environmental damage, even before there is conclusive scientific evidence that damage will occur'.

APPENDIX 4 : ECONOMIC APPRAISAL AND THE ASSESSMENT OF INFORMATION IN ENVIRONMENTAL STATEMENTS

ECONOMIC APPRAISAL AND THE ASSESSMENT OF INFORMATION IN ENVIRONMENTAL STATEMENTS

Introduction

1.1 This appendix considers the integration of environmental information into the decision-making process, specifically with reference to the evaluation of environmental costs and benefits alongside those of a purely economic nature. All references are listed in the bibliography (**Appendix 1**).

1.2 This is not a new task for local planning authorities, for whom a principal function is to manage land-use planning so that economic, social, environmental and welfare considerations are optimised. Planning involves assessing development projects from the standpoint of both the local community and society as a whole. Inherent in this process, a local authority must consider a new development in terms of the benefits and/or losses it brings to individuals. In this context 'social' refers to the aggregate of individuals' welfare.

1.3 Before discussing the potential contribution that monetary valuation can make to incorporating environmental information into decision-making, it is firstly necessary to explain two important distinctions. These are between price and value, and between financial and economic appraisal.

Price and Value

1.4 Some of the earliest economic commentators recognised that the price of a good and its value need not be the same. Adam Smith, writing more than 200 years ago, noted the extreme disparity between the (very low) price of water and its (very high) value. The use of environmental goods and services is commonly dictated by their price rather than their value. A contemporary example is the disposal of sewage sludge into the North Sea. In price terms, the sea seemed to provide a virtually free waste disposal service, but its value is now becoming apparent with the phasing out of this disposal route and the attendant requirement for costly alternatives (eg. incineration or landfill).

1.5 Such additional costs serve to show that examining the market price of alternative goods is one route by which environmental goods can be priced. This approach belongs to a wider group of evaluation techniques which attempt to put monetary assessments upon environmental goods and services by reference to other goods which do have market prices. Such approaches are termed 'pricing' techniques. These approaches are relatively simple to operate and can provide an adequate evaluation in certain circumstances. For example, a policy which reduced atmospheric discharges of sulphur oxides would, taking account of any scientific uncertainty about the impact of the pollutants have predictable effects on damage to buildings, loss rates of planted timber, and the incidence of respiratory disorders. Certain components or consequences of these effects are amenable to the attribution of (relatively) unambiguous monetary values: building restoration costs saved, timber marketed and medical costs foregone (Bowers, 1993).

1.6 In many situations these pricing techniques can only be interpreted as valid measures of economic value where environmental standards are available, as these provide an appropriate basis on which to justify the costs being incurred. For example, Pearce (DoE, 1991) notes that the use of replacement costs to measure pollution damage would only be correct if it was argued that the remedial work had to be undertaken because of some kind of environmental regulation, such as a water quality standard.

Similarly, where no sustained environmental standards exist, Bowers argues (1993) that the replacement cost approach is inappropriate for valuing the environment.

1.7 However, pricing techniques can considerably underestimate the full value of a good and may therefore still allow considerable resource misallocation. Their application thus demands careful judgment. Economic theory indicates that the full value of a good can only be determined by examining the demand for it. This involves estimating a natural resource's demand curve, showing how varying prices affect an individual's usage of the good.

1.8 The demand curve for environmental goods is calculated on the same principles as that for 'normal' marketed goods, such as bread or cars. This involves identifying people's preferences for environmental resources through their willingness to pay (or to incur costs) to secure some of its services (eg. cleaner air, or a day of outdoor recreation). Each individual's preferences (or willingness to pay) will differ. Since the aim is to define what is socially desirable, individuals' willingness to pay (WTP) are aggregated to secure a total WTP. This amount provides a monetary measure of preferences.

1.9 The 'demand curve' approach enables the difference between the market price of a good and its value to be measured. A range of demand curve techniques have been developed (contingent valuation method, travel cost, hedonic pricing). They are of particular relevance where a large disparity between price and value seems likely. In such instances the use of pricing techniques would be inadvisable since they would tend to underestimate the value of the good being appraised. However, in practice, the demand curve approaches are generally more complex than the alternative price techniques.

Financial and Economic Appraisal

1.10 In an 'economic appraisal' of a development proposal the total costs to society of the development would be compared to the benefits which the development will provide to society. This appraisal technique is known as Cost Benefit Analysis (CBA) (see paras 1.18 - 1.20). Such an economic appraisal differs significantly from that undertaken by a private developer, which is primarily aimed at assessing whether the proposed project is likely to yield sufficient profit to justify the necessary investment. This is termed a 'financial appraisal' and focuses upon the costs and benefits to the private company and so may ignore, for example, pollution damage (costs) which fall upon society (eg respiratory disease due to air pollution). In such a case the financial appraisal will show a different 'net benefit' (benefits minus costs) than the economic appraisal (the social value of the project).

The Appraisal Framework

1.11 Before considering how a financial appraisal can be modified to become an economic appraisal, it is necessary to examine in more detail the costs and benefits which can arise from a development. To illustrate this, the differences between the financial appraisal of a project and economic appraisal of the same project are considered in the context of a worked example, i.e. a hypothetical forestry company's proposal to plant trees on an area of land.

1.12 **Figure 1.1** illustrates various frameworks for appraising the proposed afforestation project. The project yields both benefit and cost items. These items can in turn be subdivided as either 'internal' (those which accrue to the private company) or 'external' (those which flow to the rest of society).

Figure 1.1: An extended cost benefit framework for appraising a proposed forest

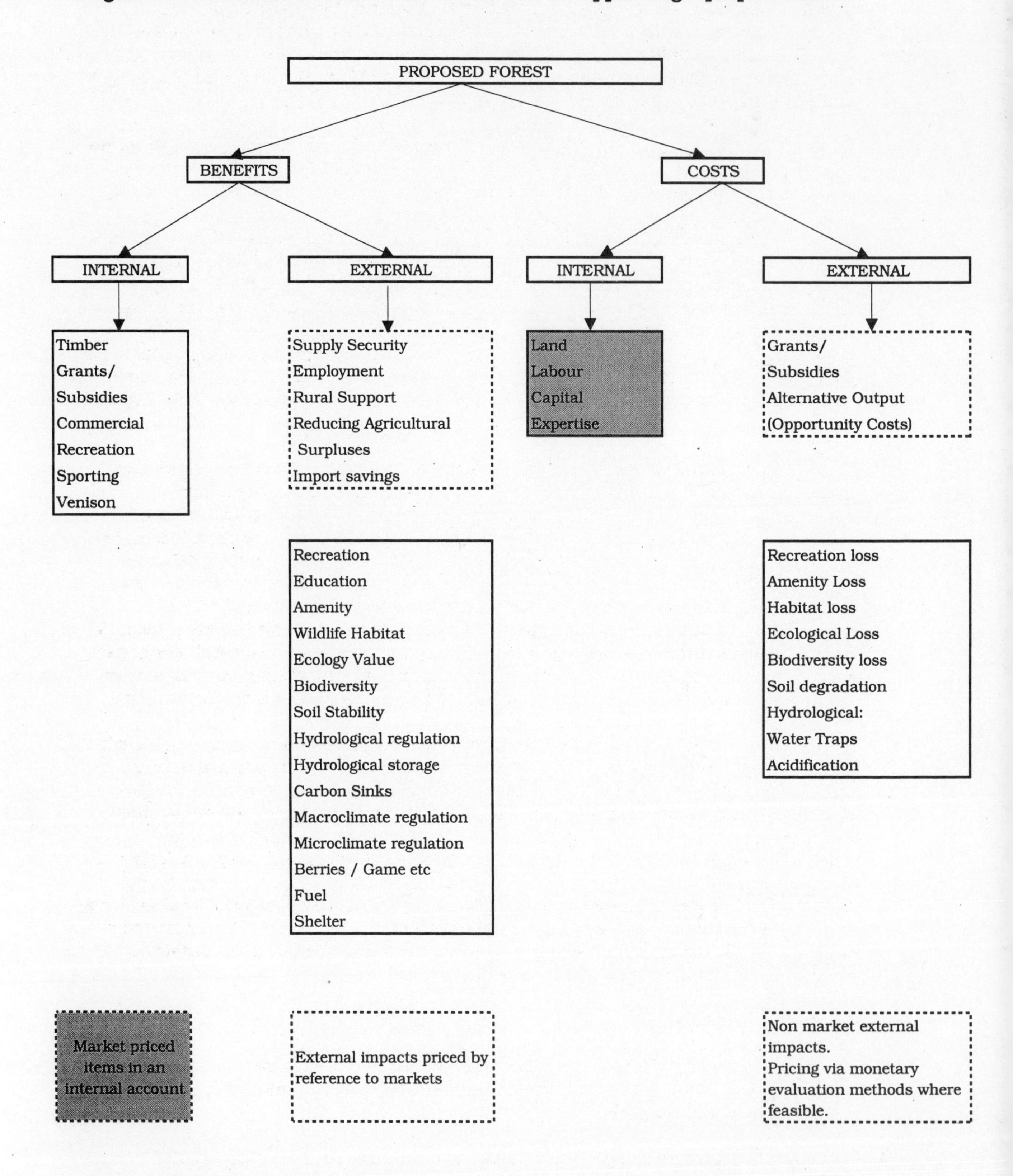

Note: Here `internal' refers to the private forestry company's costs and benefits, while `external' refers to costs and benefits accruing directly to society.

Source: Bateman (1992a)

1.13 The private forestry company will want to know if it will make a profit from the woodland. To this end it conducts a 'financial appraisal', comparing the internal benefits it receives with the internal costs it has to pay for (shaded boxes in **Figure 1.1**). If internal benefits outweigh internal costs to the extent that the company would not make more money from investing elsewhere, then it will proceed with the project and start planting trees.

1.14 All types of CBA are concerned with the wider social value of a project and so consider both the internal and external benefits and costs listed in **Figure 1.1**. A primary question for CBA analysts is how are these items to be evaluated?

1.15 In a CBA, the internal costs and benefits of a project (shaded boxes) can usually be evaluated by reference to their market prices. Some of the external items are also related to market prices (the broken line unshaded boxes of **Figure 1.1**). For example, if the land was previously being used to grow crops then planting trees will mean that these crops are lost. This is a loss to society, ie. an external cost, which can be evaluated by looking at the market price of the lost crops. However, many items do not have market prices (eg. amenity, recreation, wildlife habitat, etc) and cannot be readily valued by such an approach (those items in the solid-line unshaded boxes in **Figure 1.1**).

1.16 Until recently many CBA studies had no method for placing prices on (i.e monetising) externalities which have no connection with market prices. As many of the environmental externalities of a project fall into this category (benefits such as the creation of recreation, amenity, wildlife habitat etc., and costs such as pollution, loss of habitat, etc), this meant that the money sums resulting from such CBAs were under- representative with respect to environmental impacts. This in turn leads to decisions being taken on the basis of, at best, imperfect, and at worst, biased information. In practice the traditional response to such problems has often been to include some written qualitative environmental assessment of the project along with the CBA. However, because decision-makers tend to work with monetary measures, these non-monetary assessments have tended to be overshadowed by the former.

1.17 In addition to environmental costs and benefits there are economic externalities which require consideration by the planning authority. For example, a proposal to build an out-of-town shopping centre may cause the variety of local shops to fall, with the remaining ones incurring higher unit prices, which in turn may disadvantage those local residents who cannot afford to travel to the new centre. These economic externalities can be taken into account by the local planning authority in addition to the environmental impacts of this development, such as the loss of recreational / amenity land or a detrimental impact on landscape. It is the task of the local authority to take both economic and environmental externalities into account in arriving at a decision. While recognising this interplay, this study is confined to the evaluation of environmental externalities within project appraisal.

Cost Benefit Analysis

1.18 As noted previously, CBA is concerned with assessing the wider social value of a project and so requires the systematic identification and evaluation of internal and external costs and benefits.

1.19 Although this procedure provides a means of comparing the advantages and disadvantages of a project, criticism of the approach has centred around its inability to measure, and thus give due consideration to, environmental externalities. In practice two major criticisms raised in relation to the traditional CBA approach have been that:

171733 F

i) because conventional CBA only assesses items in monetary terms, the environmental impacts of a project are often considered in qualitative non-monetary units which, in turn, may not be given equal weight by decision-makers; and

ii) conventional CBA does not include a sustainability criterion, i.e. it does not ensure the conservation of natural resources over generations.

1.20 The first point is arguably only of relevance if the environmental externalities of a project are diverse and pull in different directions. For example, if the overall environmental impact of a development can be readily proved (through non-monetary, qualitative appraisal) to be positive and beneficial, then the planning authority may not gain any benefit from considering the results of a traditional CBA. It is often the case, however, that a development will generate an array of disparate impacts which the decision-maker is required to trade-off against one another. It follows that if there was a common unit in which to measure impacts, the difficult task of comparing them might be simplified. The Standing Advisory Committee on Trunk Road Assessment (SACTRA 1992) in considering the environmental assessment of road schemes, regarded such use of monetary valuation (in appropriate situations) as offering "a greater degree of consistency between separate decisions affecting a similar subject matter" (SACTRA '92, Chap. 13, S 13.05).

Standard CBA and the Requirement for Extended CBA

1.21 Standard CBA, i.e. that confined to a narrow financial perspective which largely excludes external costs and benefits, in particular those of an environmental nature, has a limited role to play in improving the assessment of environmental information by planning authorities. Planning authorities are not concerned with the question of whether or not a (private) development is financially viable or how profitable it is. Their interests lie in the wider costs and benefits of the scheme, which could be environmental or economic. For example, will the external environmental costs of a proposed development be offset by external environmental benefits? Can conditions be imposed on the developer to ensure a net environmental benefit is obtained? What is the scope for negotiating planning gain? It is in this context that a broader assessment, termed an extended CBA, which does take account of environmental externalities, could play a useful role in evaluating environmental information accompanying planning applications.

1.22 The monetary evaluation of environmental assets and impacts under the umbrella of an extended CBA can be achieved via a range of specific evaluation techniques. These include the contingent valuation technique, travel cost, hedonic pricing, the shadow-project approach and dose-response technique. These techniques are returned to later, and their potential application to a case study is also addressed in **Appendix 5**.

1.23 Before this, however, the potential application for extended CBA and the implications of its usage, in the context of local authority planning decisions, needs to be addressed in broader terms. As noted previously, if the environmental impacts of a project are not transposed into monetary equivalents then decision-makers may give them less attention than they might otherwise merit. Over a period of time this could lead to progressive environmental degradation. Extended CBA, when applied in appropriate situations, offers a potential means of alleviating this problem as it explicitly draws the environmental impacts of development into project appraisal.

1.24 EA addresses the environmental externalities of developments, which underlie the need for the full or extended CBA. It can highlight environmental impacts and provide a starting point for action either to avoid or to reduce them. But because EA within the planning system is essentially project specific, it does not necessarily guarantee the optimal solution for the environment.

1.25 As noted by Thomas et al 1991, "current environmental assessment arrangements emphasise mitigation of adverse environmental effects rather than consideration of alternatives which would achieve a reduced environmental impact or environmental enhancement". From an environmental economic perspective this is significant, as a broader approach would allow for an enhanced appreciation of a wider range of environmental costs and benefits associated with a selection of options relevant to a development. Specifically, if EA were more strategic in approach, and accommodated the consideration of alternatives (locations or project options) in greater detail, then it would be better equipped to anticipate and respond to environmental problems which might otherwise become acute at the individual project level.

1.26 In addition, the quality of environmental statements and accuracy of impact prediction is variable (Lee, 1993, Bisset, 1984). This is significant with respect to the potential application of monetary evaluation techniques - if baseline information on environmental impacts is inaccurate, or of poor quality, this precludes the successful translation of environmental impacts into monetary terms. Variability in the quality of environmental statements can lead to the mistrust of the information contained within them. Consequently, potentially significant environmental information concerning specific projects could be rejected by planning authorities, impeding the decision-making process. Similarly, lack of sufficient environmental information will lead to sub-optimal decisions being made, whether or not environmental monetary evaluation techniques are employed, and the use of such techniques cannot compensate for a lack of basic environmental information.

1.27 Research into post-development audits to investigate the accuracy of environmental impact predictions has led to recommendations for potential extensions of EA practice (Bisset 1984). This research recommended that EA should be "more flexible, involve rigorous hypothesis testing and should not only be concerned with impact prediction, but also with impact management once a development becomes operative". EA therefore, would become an iterative, cyclical process involving prediction, monitoring, auditing and feedback of results to project management.

1.28 An extension of EA in this way could in time become an integral component of the planning process. Decisions could then be made more firmly on the basis of the environmental information provided by the EA. If the environmental information per se were adequate for decision making, then monetary evaluation might be unnecessary, although there might still be occasions where the techniques could make a useful contribution, eg. balancing in a common unit (ie monetary) disparate impacts attributable to different project options, or if specific environmental assets which are particularly amenable to environmental economic assessment were affected.

1.29 For example, where people have preferences for changes in specific environmental assets, then an 'extended' EA analysis, whilst providing a better base for quantification of impacts, may not reveal these preferences. In such a case monetary valuation might provide a useful decision making aid. The application of environmental economic techniques for public funded development is already accepted in practice. This is because the money invested in public works, such as a major trunk road or flood defence schemes, is largely derived from taxation, and must be shown to be economically efficient on a national basis. The challenge is to see how this approach could be applied to private sector developments which are subject to planning authority control.

1.30 Another reason why EA, in its present project focussed form, may fail to secure the optimum use of natural resources is that it may result in cumulative environmental impacts attributable to a number of individual developments being ignored. For example, within a given geographical area, woodland sites of sub SSSI standard, but still of local conservation and recreational value, could be subject to ongoing degradation by a range of developments. Each loss may be insignificant in its own right, but over the longer term the cumulative loss could become substantial. A strategic EA approach would provide a means for addressing this, and as woodland sites often have a tangible environmental economic perspective, economic evaluation would appropriately reinforce the EA based approach.

Monetary Evaluation Techniques

1.31 Advances in monetary evaluation have been acknowledged by the Government in recent policy setting and project evaluation guides. Relevant examples include the DoE's 'Policy Appraisal and the Environment' (1991) and the Ministry of Agriculture, Fisheries and Food's Flood and Coastal Defence Project Appraisal Guidance Notes (1993). These set out the concept of incorporating the monetary value of environmental externalities (natural resources, environmental impacts) into both public and private sector decision making to encourage environmentally sustainable development.

1.32 **Figure 1.2** identifies the various monetary evaluation techniques currently available. As mentioned in paragraph 1.7, pricing techniques can considerably underestimate the full value of a good and may therefore result in resource misallocation. Conversely, demand curve approaches are more likely to take account of the full value of a good as they incorporate the actual demand for it. The principal demand curve approaches are discussed in more detail below.

Figure 1.2 - Monetary Evaluation Techniques for the Valuation and Pricing of Environmental Resources.

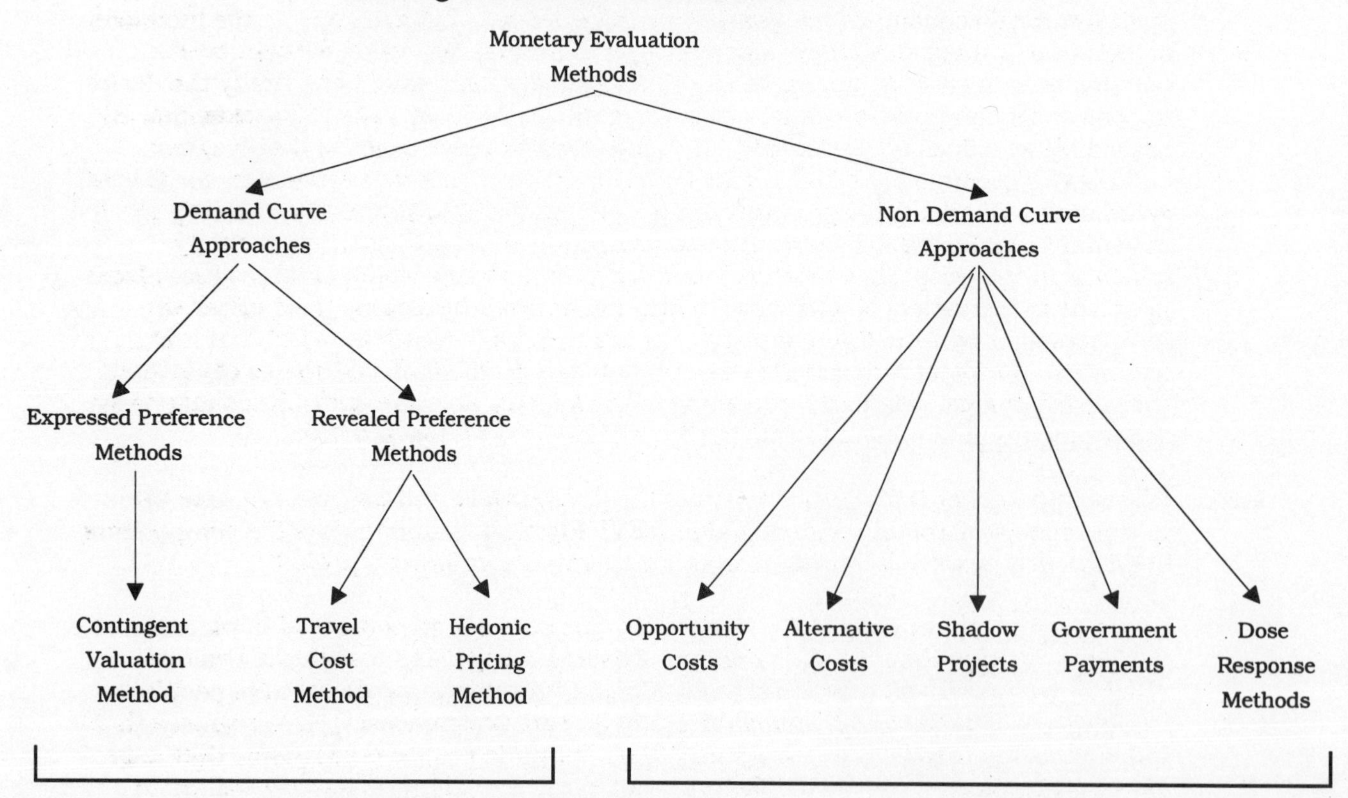

Source: Adapted from Bateman (1992)

Demand Curve 'Valuation' Approaches

1.33 The following methods all centre on the estimation of a demand curve for the environmental good in question. As such they provide true valuations rather than simple 'pricing'. **Figure 1.2** indicates that there are two basic types of demand curve evaluation methods; firstly, demand can be measured by examining individuals stated ('*expressed*') preferences for the environmental good (elicited via questionnaires); secondly, demand can be *revealed* by examining individuals purchases of market priced goods which are required in order to enjoy associated environmental goods. Before considering these valuation methods in detail, we need to consider the more fundamental question of what is meant when we say that something is of value.

What is Value?

1.34 Traditional economic theory sees value as a 'utilitarian' concept. Here items only have value because of the functions or services they provide for humans, ie. values in use or, more simply, use-values. 'Primary' use values include the intended objective of the project, for example, in appraising the forestry project discussed earlier, a 'primary' benefit would be the timber value of the forest. 'Secondary' use values include the knock-on uses generated by the project. In the forestry example this would include the value of the employment and recreation created.

1.35 An extension to utilitarian use value is the notion of an 'option' value (Bishop, 1982). This recognises that individuals who do not presently use a resource may still value the option of using it in the future. For example, they may be willing to subscribe to the maintenance of local recreational parks, which they have not previously visited but may wish to use in the future.

1.36 A major difference between use-based utilitarian value theory and that espoused by environmental economists (and underpinning extended CBA analysis), is the inclusion of 'non- use' values, defined as something an individual may value without ever personally using it or intending to use it. Two subdivisions exist here; firstly the desire to pass assets on to future generations represents a 'bequest' value. For example, an individual who does not personally enjoy forest walks may recognise that his/her children (or others) may feel otherwise, and so wish to preserve such forests for future generations. Such a concept is still based upon human use although not use by the individual expressing the value. The second non-use category is fundamentally different in this respect; 'existence' values refer to the value which an individual places upon the preservation of some asset which might never be directly used either by him/herself or by future generations. For example, individuals may feel that it is important to protect Antarctica from pollution and exploitation regardless of whether they visit the area. Similarly, returning to the forestry example, individuals may value the preservation of remote forests for the wildlife habitat they provide.

1.37 Pearce and Turner (1990) bring together these subdivisions of use and non-use value in the concept of Total Economic Value (TEV). **Figure 1.3** summarises the components of TEV together with illustrations from an appraisal of a forestry project.

1.38 Moving from utilitarian to TEV concepts of value allows for a considerable widening of people's attitudes towards the valuation of assets, recognising that human values reflect a complexity of underlying motivations. Utilitarian value theory sees people as being basically motivated by individual gain (private preferences). TEV recognises that individuals are members of society and that, alongside private preferences, they may also have altruistic motives (public preferences). These motivations may well be in conflict. As an example, in the case of planning for transport demand, individual private preferences may highlight the value of using cars for transport, irrespective of the damage caused to the environment. However, public preferences may be influenced by bequest and existence values to prefer the expansion of lower pollution public transport systems.

Figure 1.3 - An Example of Total Economic Value (TEV)

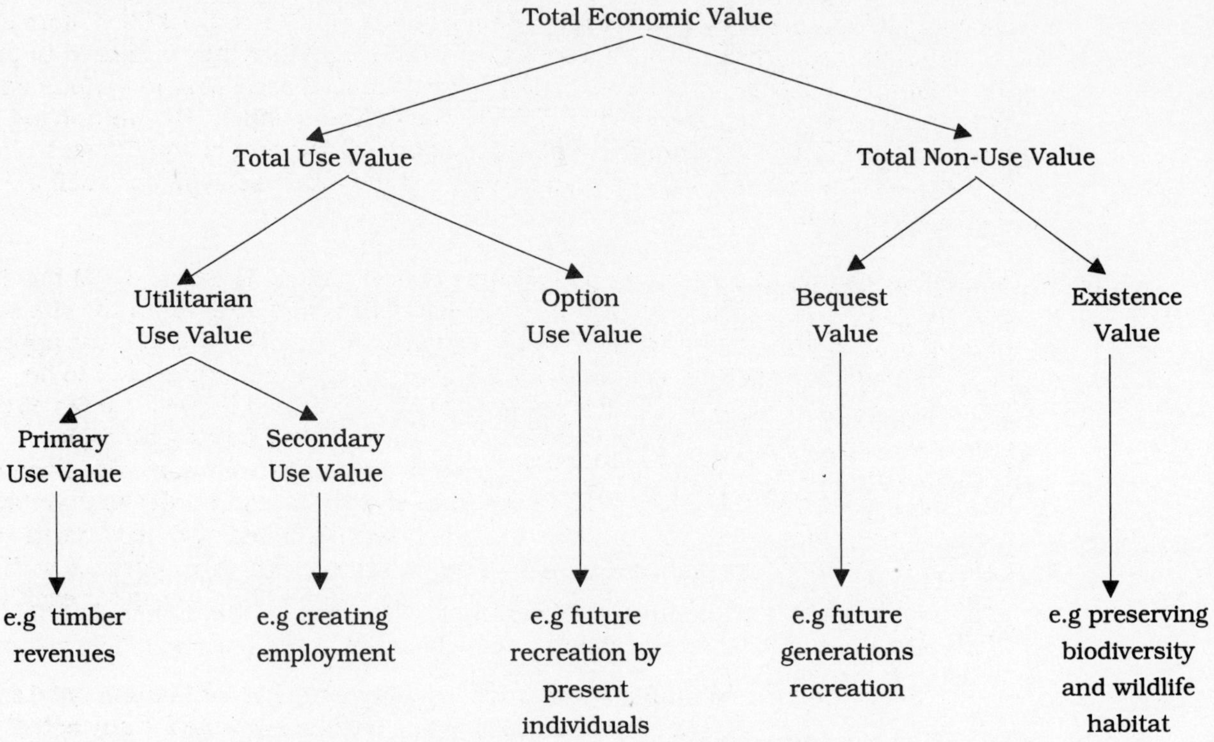

- Total Economic Value
  - Total Use Value
    - Utilitarian Use Value
      - Primary Use Value → e.g timber revenues
      - Secondary Use Value → e.g creating employment
    - Option Use Value → e.g future recreation by present individuals
  - Total Non-Use Value
    - Bequest Value → e.g future generations recreation
    - Existence Value → e.g preserving biodiversity and wildlife habitat

The various components of TEV may therefore be in conflict. Using transport as an example, the development of an expanded trunk road system may be associated with positive use-value (benefits) but negative existence / bequest values (costs). Environmental economists are therefore concerned to extend the conventional CBA framework so as to capture both the use and non-use values of environmental goods. This is the problem addressed by 'valuation' techniques.

1.39 Valuation techniques refer to people as the arbiters of value. Because of this, such methods are termed 'preference' methods and two basic variants prevail.

(i) Expressed preference methods:

1.40 These directly ask people about their valuation of environmental goods. The predominant method amongst these is the Contingent Valuation Method (CVM). Because people are both individuals and members of society, their stated evaluations are likely to be a mixture of individual (private) and public preferences regarding environmental assets or externalities.

(ii) Revealed preference methods:

1.41 These ascertain individuals' valuations of environmental assets by observing their purchases of market priced goods which are necessary to enjoy the environmental good in question (eg. purchasing petrol so that a day trip to the country can be enjoyed). These methods include the 'Travel Cost Method' (TCM) and the 'Hedonic Price Method' (HPM).

(i) Expressed Preference Methods

The Contingent Valuation Method (CVM)

1.42 This method obtains environmental evaluations by using surveys to ask people what they are willing to pay (WTP) for an environmental benefit or what they are willing to accept (WTA) for its loss. In a recent example (Willis and Garrod, 1993) visitors and residents in the Yorkshire Dales were asked to state whether they preferred to protect the existing landscape or to have the landscape changed according to various options. Respondents were then asked their WTP for their chosen option. By multiplying together the WTP for a particular landscape option by the proportion of users choosing that option it is possible to obtain an estimate of the total use value for such a landscape option.

1.43 Surveying visitors at a site gives an estimate of user values. However, CVM has the advantage that non-users can also be interviewed at points away from the site so, for that option, bequest and existence values can also be incorporated into the appraisal, ie. CVM studies, in theory, enable the full TEV of an environmental asset to be estimated. A good example of such an application is provided by Desvousges et al (1987) in their study of the Monogahela River, a major river flowing through Pennsylvania, USA. Analysts asked a representative sample of households from the local area what would be their WTP in the form of extra taxes in order to maintain or increase the water quality of the river. The analysts conducted several variants of the CVM survey. In one variant, households were presented with three possible water quality scenarios and simply asked how much they were willing to pay for each (see **Table 1.1**).

Scenario 1 : Maintain current river quality (suitable for boating only) rather than allow it to decline to a level unsuitable for any activity (including boating);

Scenario 2 : Improve the water quality from a level suitable for boating to a quality where fishing could take place;

Scenario 3 : Further improve water quality from fishable to swimmable.

Table 1.1 - Willingness to Pay for River Quality Scenarios

Water Quality Scenario	Average WTP of whole sample ($)	Average WTP of users group ($)	Average WTP of non-users group ($)
Maintain boatable river quality	25.50	45.30	14.20
Improve from boatable to fishable quality	17.60	31.30	10.80
Improve from fishable to swimmable quality	12.40	20.20	8.50

Source: Abstracted from Desvousges at al (1987)

1.44 A number of interesting conclusions can be drawn from these results. Considering the results for the whole sample, it can be seen that the stated WTP sums draw out a conventional demand curve for water quality, (ie people are prepared to pay a relatively high amount for an initial basic level of quality, but they are prepared to pay progressively less for improvements to that water quality).

1.45 The total user or non-user benefit value of a specific improvement can be estimated by multiplying the relevant household value by the number of households in either group (users and non-users). This benefit can then be compared against the cost of achieving such a quality improvement to see if it is worthwhile.

1.46 It is important to note that the WTP of non-users is not zero. This is because such households, while not personally wishing to visit the river, nevertheless do value its continued existence and even upgrading so that others can enjoy its benefits. As indicated previously, this non-use, 'existence' value derives from individual's altruistic 'public preferences'. This demonstrates that concentrating upon individuals' 'private preferences', as indicated in the prices of marketed goods, does not always fully capture the entire range of values which people have for goods, particularly those provided by the environment.

(ii) Revealed Preference Methods

The Travel Cost Method (TCM)

1.47 TCM uses the costs incurred by individuals travelling to a site as a proxy for its recreation value, assuming for example that each visitor to a country park values the park to be at least as great as the costs incurred by them in visiting the park, including transport costs, entrance fees and the value of their time. In other words, values are revealed from individuals' purchases of marketed goods. By surveying visitors to a site and asking them their travel costs and frequency of visits, it is possible to map out the demand curve for individuals, which can then be aggregated to the site level. **Table 1.2** shows the results from a TCM study of the recreational value of visits to UK forests (Willis and Benson, 1989).It shows the per visit value of visiting the forest (column 2 where column 1 is the forest name). Multiplying this value by the number of visits per year (column 3) gives the total recreational value of each forest (column 4).

Table 1.2 - Value of Recreational Visits to Four UK Forests (1987 prices)

Forest	Recreation value per visitor	Annual visitor numbers	Total recreational value	Area (ha)	Recreation value (£ per ha)
Thetford	2.51	102,000	256,020	20,000	12.80
Hamsterley	1.69	122,000	206,180	2,086	98.84
Clattering-shaws	2.27	32,000	72,640	5,870	12.37
Symonds Yat	2.11	158,000	333,380	1,440	231.51

Source: Willis and Benson (1989)

1.48 Dividing the total recreational value of each forest by forest size (column 5) allows one to calculate the recreational value per hectare of each forest (column 6). Mapping these values allows examination of where recreational values are highest and such information can be used to inform the planning and management of recreational resources.

1.49 Unlike CVM, the TCM can only estimate the user value of those people visiting the site. Option values and non-user (bequest and existence) values are not captured and therefore TEV cannot be estimated. However, user recreation values are often significant and the TCM has received worldwide application.

The Hedonic Price Method (HPM)

1.50 The HPM is also a revealed preference method. It uses information revealed by purchasing decisions to estimate the monetary value of environmental goods which do not have market prices. The most widely used case is that of house prices. Because these vary partly in accordance with environmental quality they can be used to estimate the monetary value of environmental attributes such as landscape amenity, noise and air quality.

1.51 With the HPM, the price paid for the environmental good is estimated by statistical analysis of variations in house price in relation to changes in the amount of the environmental good in question. For example, if one was interested in estimating the amenity value of broadleaved trees, one would look at how house price varied according to the number of broadleaved trees nearby as well as other relevant explanatory variables such as structural characteristics (eg. house size, number of rooms), neighbourhood characteristics (eg. accessibility to workplaces). An analysis of the variables should help to isolate the part of the house price which was paid in respect of the presence (or absence) of broadleaved trees near the house; ie. the 'price' of broadleaves. A demand curve could then be constructed to show the amenity value which broadleaved trees create.

1.52 Garrod and Willis (1992) undertook such a study of the amenity value of broadleaved trees in Britain. They found a significantly higher price being paid for houses with views of broadleaved trees, and duly estimated the value for a variety of scenarios such as planting broadleaved trees around houses or replacing conifers with broadleaved trees.

The issue of equity

1.53 The concept of CBA is based on the principles of encouraging overall economic efficiency and 'value for money' in a social context. But local authorities are also concerned with the *incidence* of external costs and benefits (ie. which costs and benefits affect certain individuals, and by how much). If this incidence were known, the authorities could debate how to trade-off a development option with the greatest net benefits against less economically 'efficient' options which will provide a more equitable distribution of costs and benefits. Unfortunately there are no firm criteria on which to base this decision (Lichfield, 1990).

1.54 In response to this uncertainty two options remain:

(i) start by assessing which is the most efficient development option then consider what is lost in order for a more equitable decision to be taken; or

(ii) start by calculating which is the most equitable development option and then consider the appropriate economic costs involved in choosing others.

Both these forms of project appraisal require more all encompassing decision making methods along the lines of those described in **Appendix 3**, which in turn are subject to their own shortcomings and practical limitations.

Conclusions

1.55 There is a case for additional guidance pertaining to environmental economic evaluation relative to decisions which have to be made by local planning authorities. In selected instances monetary evaluation could have a useful role to play. This is because the use of monetary values provides a common and comprehensible measuring rod that enables different options and environmental variables to be compared against each other. This may at least give a relative if not absolute indication of their true value.

1.56 However, it is unrealistic to assume that planning authorities could use the various techniques described, within a broader extended CBA framework, to evaluate every development application they have to consider. Generally, the approach would prove too detailed, complex and expensive to incorporate into the planning process on a routine basis.

APPENDIX 5 : STUDY OF THE POSSIBLE APPLICABILITY OF MONETARY EVALUATION TECHNIQUES TO THE ENVIRONMENTAL STATEMENT FOR A PROPOSED AIRPORT

Contents

STUDY OF THE POSSIBLE APPLICABILITY OF MONETARY EVALUATION TECHNIQUES TO THE ENVIRONMENTAL STATEMENT FOR A PROPOSED AIRPORT

1. INTRODUCTION AND BACKGROUND TO THE PLANNING APPLICATION

The primary objective of the case study is to consider the scope for applying economic evaluation to the environmental impacts arising from a proposed airport and thus illustrate the potential role of monetary evaluation techniques in assessing environmental information presented in an Environmental Statement (ES).

The proposed airport site comprises a derelict steel works, on-going opencast coal mining, and some open land comprising part of a golf course and areas of public open space. The open land is designated green belt. The site therefore appears to include both areas which may benefit from the development, and others, such as the open land which includes remnants of ancient woodland, likely to be adversely affected by the airport.

The economic benefits of an airport in this region were first investigated by the County Council which examined around thirty possible sites throughout the county. Subsequently, the respective city councils, chambers of commerce and the regional branch of the CBI formed an Airport Working Party to progress the issue. Consultants were commissioned to assess the viability of an airport and the relative merits of the preferred two sites. The site with the greatest economic development and job creation potential was selected, and this choice was further endorsed in a consultancy study on the economy of the region. The local planning authorities involved prepared a joint planning brief setting the proposed airport in the context of a wider economic development strategy.

The first application for outline planning permission for the site was made by the joint planning authorities in 1988. An ES accompanied the application, the main focus of which was to assess the effect of increased noise levels upon the local community, to detail sound insulation works and costings and consider the effects of de-icing agents upon local rivers. Air pollution was also briefly considered. A second application, with accompanying ES, was submitted in 1990, following agreement that the opencast contractor should take forward the airport proposal once extraction was completed. The ES was prepared by external consultants and provided additional information.

The ES notes that the venture will yield positive economic benefits for the surrounding region. In particular, it reports that the airport will be a significant employment generator, both through jobs directly associated with its construction and operation, and through new employment opportunities at an associated industrial estate, to be redeveloped adjacent to the airport.

The advantages of proceeding with a given project can often be clarified by comparing the benefits and costs of the proposed development against a "do-nothing" option. Comparison against the do-nothing option in this case, however, is complicated by the fact that the organisation which expressed an interest in winning coal from the site offered to make a substantial contribution to the construction of the airport if that was the preferred after-use. Throughout the ES, therefore, the existing coal extraction activity and proposed airport development are compared against one another, such that specific benefits of one scheme (ie. restoration of land following cessation of mining) are transferred to the other (ie. the airport) and visa versa. For example, the ES accrues the environmental benefits of restoring the mined area to the airport development. However, this area would be restored upon exhaustion of the coal reserves regardless of whether the airport development proceeds. A first step of project appraisal is therefore to clearly define the actual benefits and costs, direct and indirect, that would accue with and without the proposed airport development proceeding.

The advantages and disadvantages of the range of techniques (described in **Appendix 4** to the Research Report) that economists have used to gauge the economic importance of environmental assets affected by developments are considered. In particular, the case study seeks to address the following issues:

- the importance within cost benefit analysis (CBA) of clearly defining the range of project options, and their boundaries, and how this could affect the requirements of an ES;

- whether attributing a monetary value to conservation benefits foregone to development can be introduced to the decision-making process;

In addition, recommendations from the Standing Advisory Committee on Trunk Road Assessment (SACTRA, 1992) study, on the evaluation of the environmental effects of major road schemes are considered. While the SACTRA report was orientated towards roads, its general approach is applicable to other types of development. In particular the recommendation to itemise and cost all noted mitigation actions, whether or not they are applied in practice, is assessed in the context of the proposed airport.

2. COST BENEFIT ANALYSIS

The proponents of cost benefit analysis argue that ESs and planning procedures inadequately evaluate environmental resources/impacts because they assess items in different units to those of the economic effects of a development. The latter, it is argued, often receive the most attention in the project appraisal process, so ignoring or potentially reducing the significance of environmental impacts.

In the past, the inability of conventional CBA to evaluate environmental impacts on an equal basis with the financial implications of a project has led to criticism of its use. However, over the past two decades a range of monetary evaluation techniques have been developed which attribute monetary values to non-market environmental goods and services.

The application of these techniques has improved the effectiveness of CBA, hence the term 'extended' cost benefit analysis. This project appraisal method involves identifying, in monetary terms, all the costs and benefits of a project in order to determine whether a net gain or loss, arises from the project.

The environmental impacts arising from the airport can be divided between those arising from its construction (e.g. loss of woodland), and those pertaining to its use (e.g. air pollution). The identification and categorisation of impacts, which is essentially the environmental assessment process, is required before monetary evaluation can be conducted. The main impacts associated directly with the airport which are potentially amenable to monetary evaluation are:-

Construction Related Impacts

1. Loss of ancient and mature deciduous woodland areas (Section 4)
2. Loss of part of the golf course (Section 5)
3. Loss of green belt and associated areas of public open space (Section 6)

Airport Use Related Impacts

4. Noise pollution from aircraft and airport generated traffic } (Section 7)
5. Air pollution from aircraft and airport generated traffic }

The applicability of specific monetary evaluation techniques to these main impacts is discussed below. **It is not, however, an objective of this case-study to conduct an actual evaluation of these impacts, but to illustrate the scope for their successful application.**

3. ANCIENT AND MATURE DECIDUOUS WOODLAND

A potentially significant ecological impact of the airport is the possible loss of, or damage to, remnants of ancient and mature deciduous woodland and grassland including a short stretch of medieval wood bank. In relation to the potential application of evaluation techniques it is important to distinguish between ancient and mature woodland. Ancient woodlands are of greater environmental value since they represent a fixed stock of irreplaceable, natural resources which are typically of high conservation value. Mature woodlands, although of conservation interest, are not of the same inherent environmental value. 'New' woodland areas could be replanted which will, once the trees have matured, compensate for mature (but not ancient) woodland lost to a development. The potential to evaluate ancient woodland in monetary terms is discussed because of the habitat's significant nature conservation value in Britain.

The ES states that the proposed development site contains remnants of ancient woodland and areas of mature deciduous woodland. The report is unspecific, however, on their extent and characteristics (e.g. richness of flora and fauna). Some of these areas have already succumbed to open-cast mining operations, although the degree of this damage is not documented. The ES does not clearly distinguish the impact of the airport from that of mining operations on ancient and deciduous woodland.

However, the ES notes that "mature woodland" would be lost to the airport development and that the "surviving ancient woodland is to be integrated" with the creation of an `environmental corridor', to be located around the runway fence.

3.1 The Nature Conservation Importance of Ancient Woodland

There are only about 340,000 hectares of ancient woodland in Britain (1.25% of total land cover), but this comprises a significant wildlife habitat (Hanley et al, 1991). Although remnants of original ancient woodland remain, most have been modified by past management regimes, and so should be regarded as being only semi-natural in origin. Ancient woodlands are important not only for their stock of native trees and great age (they usually have records of their existence dating back to around 1600 to 1700AD), but also for their rich associated ground flora and their fauna.

3.2 Ancient Woodland in the Context of the Proposed Airport

The lack of information in the ES regarding the ancient woodland at the proposed airport site makes it difficult to determine the specific environmental qualities of these woodland areas, and therefore what impact the airport will have upon them. However, even if the development does not actually entail the felling of ancient woodland known to remain around this site, and manages to integrate them into the scheme, it is relevant to estimate their value in order to justify their preservation costs. This may also involve evaluating what nature conservation value would be lost even if these trees remained.

The initial ES notes that the area is listed in the Nature Conservancy Council's (NCC, whose duties in England now form part of English Nature) Invertebrate Site Register and is a Site of Special Scientific Interest. The golf course in the green belt also has woods listed in the NCC's Inventory of Ancient Woodlands and is similarly recorded as a scheduled Site of Scientific Interest. This is a county level designation, but no information is given to indicate the environmental quality of such sites, or how they are designated.

The lack of information relating to the ancient woodland makes it very difficult on the basis of the ES to ascertain either the local ecological importance of these specific woods or their importance as part of a dwindling national resource stock. Such information would be required if these woods were to be valued via monetary evaluation techniques. In particular, if the Safe Minimum Standard (SMS) approach was to be adopted it would be practically impossible to assess these assets without detailed information regarding their nature conservation status and importance. It is likely that English Nature would hold information sufficient for this task.

3.3 Measurement of the Preservation Benefits of Ancient Woodland

An initial step in the procedure for applying either valuation or pricing techniques is to identify the type of benefits which ancient woodlands provide society. This issue is considered as part of the review of techniques set out below.

3.3.1 Contingent Valuation Method

One evaluation technique which has been used for the monetary evaluation of ancient woodland in the UK is the Contingent Valuation Method (CVM). For example, this technique was used to estimate the preservation benefits arising from preventing the fragmentation of Birkham Wood, an ancient woodland under threat from the Harrogate/Knaresborough bypass (Hanley et al, 1992). The Birkham Wood study, which was undertaken for the Nature Conservancy Council by the Department of Economics at the University of Stirling, highlights the public's preference for preserving woodland environments for the benefits they provide (e.g. recreation, wildlife value), and illustrates the potential role of the CVM in evaluating such resources.

As part of the study, 1000 postal questionnaires were sent to householders in surrounding areas. Respondents were asked to rank the importance of nature conservation, landscape and informal recreation in relation to Birkham Wood. The most highly ranked reason for visiting the wood was walking, followed by a desire for peace and quiet and nature study. Further analysis revealed that the most important reason for respondents wishing to preserve the wood was for its nature conservation attributes, even if people's main use of the area was for exercise rather than to observe wildlife.

The CVM study determined, at an 8% discount rate, an annual preservation benefit for Birkham Wood of £1.03/person/year. This monetary estimate indicates the local residents' willingness to pay to preserve the integrity of the ancient woodland at Birkham Wood from the proposed bypass development. However, this study did not measure benefits that accrue to residents who do not live in the immediate vicinity of the proposed bypass. The inclusion of these values may significantly alter the benefit estimates.

Changes in the information provided to respondents in the CVM was found not to significantly influence the resulting bids. So, for example, the knowledge that the site was (or was not) of national importance did not significantly affect people's valuation of the wood.

One of the aims of the study was to attempt to justify a bypass route which would ensure that the wood remained unaffected by the road. The alternative route incurred extra construction costs of £267,000. If the preservation benefits (ie benefits which accrue to society through the knowledge that the woodland remains intact) equalled this sum, then the revised route would become economically efficient. In this instance the results of the CVM indicated that the estimated benefits were insufficient to render the detour economically viable. However, as mentioned previously, the measured benefits that accrue to people outside the Birkham Wood area were ignored in this analysis, their inclusion would undoubtedly increase the benefits estimate. These benefits may originate from people who are never likely to visit or even see the wood but still value its preservation.

Such a valuation from individuals is termed "non-use" value, and can relate to existence value (defined above) or option value, where individuals value environments (or natural resources) in the expectation that they (or, for example, their children) might one day use them (e.g. visit them for recreational use). The CVM is equipped to estimate non-use values, however, this substantially widens the usual remit of CBA. The monetary evaluation of non-use values is also a relatively complex and costly operation to undertake. It is therefore unlikely to be suitable for use as a valuation technique on a routine basis for the assessment of development proposals by planning authorities. Furthermore, the CVM study undertaken for Birkham Wood could be criticised for confining the analysis to a purely postal based questionnaire, an approach demonstrated not to secure estimations as valid as those from direct, face-to-face surveys.

3.3.2 **Travel Cost Method**

Several Travel Cost (TC) studies have been conducted to estimate the recreational value of woodlands. This technique primarily involves collecting data on numbers visiting recreational sites, and the costs they incur in reaching them (e.g.. petrol costs). The main criticisms of the technique include the issue of whether or not to include travel time as a cost or a benefit, and the problem of competing sites and multi-purpose trips. In addition, it ignores any existence value which a non-visitor may have for preserving a site. It is therefore best used as providing a minimum first estimate.

Although the monetary evaluation of the recreational use of the ancient (and mature) deciduous woodland is highly unlikely to outweigh the benefits of developing the site, the TC value represents just one of a stream of environmental benefits associated with the existing site, which in aggregate may prove to be substantial.

With regard to the proposed airport site, the recreational area subject to evaluation would have to be clearly defined. It is likely that this area would include not just the remnants of ancient woodland, but also areas of public open space, and possibly parts of the golf course (although a separate TC study could be conducted for golf club members) which would be lost to the development. It might not therefore be feasible to determine in isolation the recreational value of the surviving ancient woodland, but rather to use the TC method to estimate the recreational value of the site as a whole.

171733 G

3.3.3 **Hedonic Price Method (HPM)**

This technique could be used to estimate the amenity value of the ancient woodland. The amenity value which woodlands provide can be defined as the advantage to individuals of seeing or visiting these specific woodland environments. In the context of the HPM, this satisfaction can be evaluated through its effect on residential property prices situated nearby wooded environments. Studies have shown that house prices are affected by many factors, one of which is local environmental characteristics such as air quality or their proximity to scenic landscapes. A recent UK study showed that landscape can contribute up to 7% of total house price (Garrod & Willis, 1991). For the HPM to be applicable, there must obviously be residential properties in the vicinity of a development.

The ES noted the proximity of residential properties to the airport as "relatively distant". The nearest residential area is located 750 metres from the runway.

In relation to the remnants of ancient woodland and associated areas of public open space at the proposed development site, the HPM could in principle be employed to investigate whether these existing landscape and amenity features affect house prices. The results would provide an implicit monetary value of these specific environmental resources. A considerable amount of detailed information regarding the local housing stock would be required to conduct an HPM survey, including house price determinants such as architecture, house size, number of rooms and local facilities. In addition, a large sample population of houses would be required. Statistical analysis of the collected data can isolate the effect which a particular environmental attribute (e.g. proximity to ancient woodland) has upon the house price. It is unlikely in this instance that the HPM could isolate the importance (in monetary terms) of the remnants of ancient woodland from the other recreational aspects of the site (e.g. areas of public open space or mature deciduous woodland).

Once obtained, this value could represent an external cost associated with developing the site and could, in principle, be compared against any external benefits which residential properties may gain from the airport. For example, the ES stated that the airport would have a positive knock-on effect due to increased employment prospects for the area, commenting that it would increase land and property prices throughout the region. However, the ES did not quantify these socio-economic benefits, commenting that there is no "satisfactory method" for doing so.

In relation to the HPM, there are environmental characteristics of the area (both with and without the airport) which could be reflected in house prices. If used selectively the technique can estimate, in monetary terms, particular environmental benefits associated with the existing site, e.g.. woodland areas. The results from any previous HPM studies related to the operation of airports, may form the basis for estimating the impact of, for example, air and noise pollution. The HPM could therefore provide the decision maker with more information regarding the wider external costs and benefits associated with the airport scheme.

3.3.4 **Market Price Approaches: Ancient Woodland**

(a) Shadow-Project Approach

The shadow-project approach examines the financial costs of providing an equal alternative environmental good elsewhere. Various alternatives can be considered, such as asset reconstruction (providing an equal alternative site elsewhere); asset transplantation (moving existing habitat to a new site); and asset restoration (enhancing an existing degraded habitat elsewhere). An assessment is required to determine whether these options can adequately compensate, in ecological terms, for the damage suffered by the original site under threat of development.

In relation to the loss of remnants of ancient woodland it is likely that the only acceptable shadow-project alternative would be to replace like with like. Any other replacement would not compensate, on equal terms, for the lost value of the ancient woodland. In practical terms, however, replacement is not possible and ancient woodlands are deemed irreplaceable.

In theory, it may be possible that the remnants of ancient woodland under threat from the airport could be transplanted. This would ensure that the fixed natural stock of this particular resource is held constant. The costs incurred with this requirement could be incorporated into a CBA of the development. If the airport proposal passed the CBA `test' with these additional shadow-project costs then it could be argued that the development should proceed (only in this respect), as long as the risks associated with undertaking the shadow-project were acceptable (ie. evidence of a high rate of transplantation success). In practice, however, ancient woodlands are not amenable to transplantation (Hanley et al, 1991).

Hanley et al (1991) documents that, to date, English Nature has noted some 55 habitat transplantation projects which have been either conducted or proposed in the UK, although thus far no site is known to have been moved in its entirety (Hopkins, 1988).

(b) Mitigation Costs

The ES detailed mitigation actions to avoid the felling of ancient woodland. These include coppicing the woods near the airport security fence to "protect aircraft approach surfaces and ensure safety of flight operations". The cost of avoiding damage to these particular woodland assets implies that their value must at least equal the costs of preserving them.

Such preservation costs can be used in some instances as a proxy for the monetary value of the environmental good being protected. This estimate is likely to be much lower, however, than the full value of the woodland; the question is whether it is better than no value at all. An incorrect valuation could be potentially as misleading as no valuation, if not more so.

3.3.5 Safe Minimum Standards

The remaining ancient woodland in the UK could be considered to represent the safe minimum stock of this habitat type, with its associated flora and fauna (Hanley et al, 1991). In the context of the SMS approach, this would imply that the destruction of any ancient woodland (in part, or as a whole) takes the stock of this habitat type below acceptable levels.

The SMS doctrine is much more `rigid', in terms of preserving areas of nature conservation importance from development, than the (expanded) CBA approach. Under the latter decision-making framework there is nearly always scope for trade-offs to be made between the environment and financial benefits. However, if the SMS approach was strictly adhered to, developments would only be allowed to proceed if it satisfied the ecological rules of this approach. For example, with regard to the proposed airport, this development could only proceed within the SMS approach if it could be demonstrated that none of the ancient woodland located at the site would be affected.

4. LOSS OF PART OF THE GOLF COURSE

4.1 The ES documents the proposed airport's impact on the golf course as:

> "Earthworks related to future airport construction ... extend into the northern part of the golf course with the consequent loss of 3 holes (replacement holes are to be located to the south of the airport) and the loss of ancient woodland"

The acreage or percentage of the golf course which will be damaged or permanently lost to the development is not quantified. Neither is the location of these new holes identified, nor the ameliorative measures needed to re-design the course detailed. Consequently, the magnitude and significance of these impacts remains unclear (both in monetary and non-monetary terms).

The lack of information in the ES about the golf course has meant that some assumptions have had to be made for the purpose of the case study. A first step in evaluating the effect of the proposed airport on the golf course is to divide the impacts between the construction and operational phases of the proposed development. With regard to the construction phase, the impacts are likely to be minor and of a temporary nature. However, more significant effects regarding the long term use and quality of the golf course could arise from the operational aspects of the airport (e.g. noise from overflying aeroplanes). Before the scope for evaluating these impacts is examined, the importance of the golf course to the local community is considered.

The golf course is likely to be of importance to the local environment in many respects. Primarily, along with adjacent open areas it makes a significant contribution to the amenity value of a region which is already heavily built-up. Furthermore, it may provide a habitat link for wildlife between the city and the countryside.

The golf course also satisfies a demand for outdoor sporting and recreational pursuits. The ES documents that the golf course is a municipal facility, indicating that the general public have access to the course. However, it is unclear whether its users are predominantly local or travel considerable distances to reach the course. Information relating to the accessibility and use of this facility by locals would indicate its importance to the community.

4.2 Measurement of the Environmental Benefits of the Golf Course

In relation to the golf course, monetary evaluation techniques could, in principle, be used to estimate both the environmental benefits of the existing golf course (e.g. recreational value) and the costs incurred by this resource through the construction and operation of the airport (e.g. noise from overflying aircraft). The potential applicability of the various techniques is described below.

4.2.1 The Contingent Valuation Method (CVM)

The CVM can be used to evaluate, in a monetary form, the use and non-use benefits of environmental assets. It would therefore be possible, to conduct a CVM to investigate the preservation benefits of maintaining the golf course in its existing state. However, the application of the CVM is relatively time consuming and expensive. Where other, more immediate and far cheaper techniques exist (as applies in this case i.e. foregone golf course revenues) the CVM would not be an appropriate valuation technique. Consequently, as a general rule, just because specific valuation (or pricing) techniques are, in principle, applicable to certain environmental assets (or impacts) under assessment, this is not sufficient reason in itself to justify their actual application.

4.2.2 **The Travel Cost Method**

Valuation methods could also be used to calculate the recreational value of the golf course. For example, the travel cost method could be applied to those individuals who use the facility such that the recreational value of the course would equate to the admission fee which users may have to pay. This would include the existing fee plus any increase the users would be prepared to pay.

4.2.3 **The Hedonic Price Method**

The HPM could also, in theory, be employed to evaluate the environmental benefits associated with the course although the costs and data requirements of such an exercise would be high. However, environmental impacts measured this way typically have to be above a threshold level in order to elicit an effect on local house prices. It is possible that the golf course could generate such an effect, especially if extended to include the areas of derelict or run-down land.

4.2.4 **Market Price Approaches: The Golf Course**

(a) Loss of Golf Course Revenues

The damage caused to the golf course could be measured in terms of any lost revenues, apportioned according to the extent of the course damaged by the development. This foregone revenue could be used as a proxy monetary estimate of recreational losses suffered by the golf course as a result of the development.

(b) Mitigation Costs

According to SACTRA (1992) guidance, all mitigation actions highlighted in an EA of a development should be appraised and costed, irrespective of whether or not they are ultimately adopted. Conversely, the DoE (1991) notes that remedial work costs should only be used as a proxy measure of economic welfare costs when the mitigation costs have to be incurred in order to meet a specific environmental quality standard. It notes that :

> "for example, where there is a mandatory water quality standard the costs of achieving that standard are a proxy for the benefits of reaching that standard ... because society can be construed as having sanctioned the cost (or at least the minimum cost) of setting the standard".

A decision not to implement a particular mitigation measure implies, therefore, that society does not value the environmental resource sufficiently highly to justify the costs incurred.

Mitigation actions conducted in relation to the golf course include the redesign of part of the course actually lost to the proposed airport and rerouting of some of the public rights of way which are permanently lost to the development. These mitigation costs could be construed as reflecting **part** of the impact of the airport on existing recreational resources.

4.2.5 **Safe Minimum Standards (SMS)**

The underlying concept of the SMS approach is to preserve a minimum level of natural environmental systems and to encourage a natural and wide diversity of plants and animals. Consequently, golf courses *per se* should only be considered in this context if they are known to hold specific environmental importance, e.g. in terms of their stock of flora and fauna. The generally limited ecological interest of golf courses means that application of SMS in relation to such facilities is likely to be very limited.

5.0. LOSS OF THE GREEN BELT

5.1 The Environmental Importance of the Green Belt

Green Belts are areas of predominantly open, often agricultural, land surrounding urban settlements. Their purpose is to check the unrestricted sprawl of large built-up areas and safeguard the surrounding countryside from further encroachment, to prevent neighbouring towns from merging into one another and to preserve the special character of historic towns. Green Belts are established through development plans. Development within them is subject to tight restrictions.

Relatively little attention has been given to enhancing the recreation and amenity potential of Green Belts, despite the positive role they could play in providing access to open countryside for city dwellers (Willis et al 1992).

5.2 The Green Belt in the Context of the Proposed Airport

In relation to the proposed airport, the ES reports that the eastern part of the runway and part of the airport apron falls within the Green Belt, entailing the loss of part of some golf course land. Terminal buildings and airport facilities fall outside these boundaries.

The ES further documents that some grassland and some agricultural grassland of little ecological value will be lost within the Green Belt. It also notes that a permanent diversion of public rights of way is required on parts of the golf course where the proposed airport is to be built.

The ES reiterates the local planning authority's stance towards the nature of acceptable development within the green belt, that "development will not be permitted except in exceptional circumstances for purposes other than agriculture, forestry, recreation, cemeteries, institutions standing in large grounds and other uses appropriate to rural areas". Although the ES notes that there is no Council planning policy specifically relating to airport runway construction within the green belt, the ES notes with regard to the proposed access road to serve the airport, the local authority's requirement that "every effort will be made to minimise any adverse effect such schemes may have on agriculture, amenity and recreation areas, the landscape, and sites of natural history, geological and archaeological interest".

5.3 Valuing the Green Belt

Considerable research has been directed at valuing green belts. Robinson (1990) argues that in addition to the knock-on effect of concentrating building development within urban areas, within the green belt itself, planning restriction can significantly raise the price of both land and housing. He cites Cloke's (1983) argument that green belt policies have maintained inequalities in the rural-urban fringe by favouring middle-class commuters and high income groups as a result of high residential property prices. This has led, in some areas, to the needs of local communities (e.g. low cost housing) being ignored. In relation to land prices, because green belts act only as 'guidelines' to planners, their presence in some instances has fuelled speculative landholding, which has occasionally yielded highly profitable returns, as in the case of the deletion of part of the West Midlands Green Belt for sale as development land (Robinson 1990).

The market price of land in the green belt reflects both the land's existing productivity value (e.g. if in agricultural usage, its value can be reflected through its marketed output) and the degree of expectation regarding its future possible uses. If there is a possibility of green belt land being used for development then its value is likely to increase to reflect the potential financial returns from that development. However, without this expectation, the existing market value of the land is typically low due to the relatively poor financial returns from agriculture.

Green belt land also holds a variety of non- marketed outputs which contribute little to its marketable value because of their `zero price' status, but which are considered to be environmental benefits. These are listed by Willis and Whitby (1985) as:

> "(1) Amenity value or aesthetic benefits, from a more pleasant rural environment to neighbouring local residents, and by preserving the special character of villages.
>
> (2) Recreational value, from informal or open access non-priced recreation across the land.
>
> (3) Wildlife value."

The potential benefits associated with preserving the green belt land together with the financial return from agricultural output could be considered to represent its environmental value.

If the additional benefits of preserving the green belt were realised in monetary terms, in certain circumstances, this would decrease the potential financial returns from developing this land. The following section outlines the monetary evaluation methods which can, in theory, be applied to value the non-marketed environmental outputs of the green belt.

5.3.1 Contingent Valuation Method

The external benefits of the green belt could be evaluated using the CVM (as discussed in the above study by Willis et at, 1992). These benefits are embedded in the recreational value of the land, which provides access to open spaces for informal recreational uses. In addition, green belts provide an amenity to neighbouring residents, which can be measured with respect to changes in house prices. This change in house prices can be calculated, in theory, through the application of the CVM, and not just via the adoption of the HPM. In the context of the CVM study discussed earlier (Willis et al., 1992) it was deduced that both housing and opencast mining development in the green belt significantly reduced local house prices. This result was not unexpected; neither was the strong correlation between the size of the effect of the development on house prices and the distance of houses from the site, and the ability to see the site from the individual houses.

In the context of this particular study (Willis et al 1992), it was concluded that the agricultural output of the land in the green belt was small, as was the potential loss of recreational benefits associated with its development. However, losses in terms of amenity value to individual local residents, measured with respect to changes in house prices, were found to be substantial.

The CVM approach outlined in the Willis study, could be used to evaluate the potential effect of the proposed airport upon the recreation and amenity benefits, expressed through their effect on local house prices.

5.3.2 **Travel Cost Method**

In addition to estimating the recreational value associated with the existing area of green belt, the TCM could also be used to estimate the value of the developer's commitment to provide an area of land near to the development, to compensate for the loss of the original site. The planning authority may wish to ensure that the compensating land provides the same level of satisfaction to that afforded by the existing site. Moreover, in selecting an alternative site, provision should, in theory, be made for those existing users on low incomes who may not be able to afford to travel longer distances to a new area.

5.3.3 **Hedonic Price Method**

The HPM could be employed to value the amenity value of land in the green belt.

A basic assumption of the property value approach is that changes in the existing environmental quality of an area affect the future environmental value component of residential property prices, with the result that, with other factors remaining constant, the sale price of the property may change. Properties located within the green belt may represent a special case since their designation provides them with a degree of surety that the landscape component of their existing environment will be preserved, through planning constraints, into the future. Thus the properties within a green belt may maintain a premium price through the secured protection of certain environmental features of the area.
Furthermore, the price of any property is affected by the supply of alternative, potentially substitutable properties with equivalent characteristics. Since the building of new properties in the green belt is typically restricted, existing properties can command a higher price simply because of their relative scarcity. The reason for these increased property prices is primarily tied to the planning restriction on the land and is not specifically related to the environmental amenity value of the area. A HPM valuation of an area of green belt would have to take this into account.

Properties located outside but adjacent to the green belt are also likely to benefit from its non-marketed facilities (eg. amenity). These non-marketed facilities or environmental assets could, in principle, be evaluated using HPM. A negative effect on the value of properties at the proposed airport site which are adjacent to the green belt could be expected if there was a loss of recreational open space, or an impact on views from individual properties.

5.3.4 **Market Price Approaches: The Green Belt**

(a) Shadow-Project Evaluation of Land in the Green Belt

It would appear possible to use the Shadow-Project technique to evaluate the area of green belt land lost as a result of the airport. This would require estimating the value of an equivalent area elsewhere.

In practice careful thought would have to be given to, for example, the location of the substitute site. Social costs and benefits are usually accounted for on a national basis, eg. cost benefit analyses for Government-funded projects, such as roads or coastal defence works. Consequently, it could be argued that substitute sites could be located anywhere within the UK. However, it is obviously preferable to locate a replacement site as close as possible to the original airport site in an attempt to compensate directly those people actually affected by the development. This is clearly an issue of relevance to local residents near the proposed airport site, who make use of the area.

(b) Mitigation Costs

The ES for the proposed airport does not specify environmental mitigation actions for different areas of land. Consequently, it is not possible to distinguish environmental mitigation measures directed solely at preserving the area of green belt. This is because it is not clear whether any compensatory land provided by the developer would ever be designated as green belt.

5.3.5 Safe Minimum Standards (SMS)

The application of safe minimum standards to the loss of the green belt as a result of the proposed airport hinges on the extent to which there is a policy justification for the retention of the total area of land designated as such. PPG2 'Green Belts' emphasises the importance of protecting green belts, the implication being that any erosion of them would be a net loss to the nation. It might be argued therefore that the airport is unacceptable because the total area of green belt would fall below the national safe minimum standards.

5.3.6 The SACTRA Evaluation of Land in the Green Belt

The 1992 SACTRA guidelines for the environmental appraisal of major road schemes make specific reference to the valuation of land in the green belt.

The report argued that green belt land lost to development should be valued at its potential development value and not at its current use value (agricultural land price). The SACTRA initiative would help to reduce the potentially very large financial returns which can be made from developing land in the Green Belt, by increasing its price to that of its potential development value. In turn, this may help to alleviate the speculative landholding (in the anticipation of the relaxation of planning restrictions) and encourage developers to look either within defined urban planning boundaries or beyond the green belt.

In the context of the proposed airport, if the SACTRA evaluation is applied, then this area of land should be valued at its potential development value. It is not clear, however, how this latter valuation would be calculated. For instance, would the potential development value relate specifically to the financial returns from the proposed airport, or from development in general? This may substantially affect the `potential development' value of the site.

6. NOISE AND POLLUTION ARISING FROM THE PROPOSED AIRPORT

The proposed airport would be likely to give rise to air pollution and noise impacts during both construction and operation. These impacts are discussed in parallel since the techniques for assessing their impacts in monetary terms raise similar issues.

6.1 Noise Impacts

Despite the fact that noise was likely to be one of the more significant impacts arising from the airport, its assessment in the original ES was limited. In particular, an Environmental Health Officer noted that the initial noise estimates were based upon a minimum traffic movement prediction, and not a minimum-maximum scenario. This was found to affect significantly the predicted noise levels. The local planning authority therefore requested additional information regarding predicted noise levels.

The revised ES provided a supplementary technical report on the potential noise impacts, prepared by an aviation consultant. In relation to regulatory limits controlling noise levels, the ES stated that the following standards should be investigated:

> "Safety and noise provisions set out by the CAA and CAP 168 Licensing of Aerodromes and the Aerodrome Standards Dept. must be taken into account, as well as planning criteria set out in DoE Circular 10/73 `planning and noise'".

The revised ES concluded that the effect of the airport on surrounding developed areas would be insignificant compared to the existing high noise levels from the nearby motorway, other major highways and industrial and heavy steel industry plants. The ES described the proximity of residential areas to the airport as "relatively distant" with the nearest area being located 750 metres from the runway. However, the ES noted that approximately 123 residential properties fell within the limit at which DoE guidelines normally recommend insulation schemes for residential properties, a noise level of 40 NNI.

This external environmental effect arising from airport operation could be initially costed using monetary estimates of the noise mitigation actions itemised in the ES. In this instance, however, the evaluation of night-time disturbance costs is not applicable since a planning condition states that no aircraft movements shall take place between 22.45 hours and 06.30 hours.

6.2. Air Quality Impacts

The proposed airport development would generate atmospheric pollution via:

- airborne dust and emissions resulting from airport construction;
- aircraft emissions in air and on the ground; and
- emissions from vehicles associated with airport operation.

The ES focussed the assessment of localised air quality impacts arising from the airport. However, the geographical extent of atmospheric impacts could also be extended to the global environment, for example, in relation to emission of greenhouse gases. Although the significance of the airport's contribution to total global emissions would be exceedingly small, it is the combined emissions of all polluters which may cause climate change. In response to the potential dangers of global climate warming, considerable research effort has been directed towards determining monetary damage costs attributable to certain greenhouse gases. An average global damage cost for CO_2 has been determined of approximately $10 per tonne emitted (pers. comm. Bateman, I. J., 1993). Such a value could be applied to the proposed airport to indicate, in part, the global environmental damage costs associated with its operation.

The impact of the airport on air quality is complicated by the existing impact from opencast coal mining operations at the site. The ES recognises that current air quality at the site is poor, due, in part, to these activities. However it does not differentiate air pollution attributable to mining activity from that attributable to vehicle emissions from the nearby motorway or emissions from industrial activities located near to the site. The development of the airport would effectively end air pollution from the opencast mining, but would replace it with its own specific forms of air pollution. The important question is whether this represents an environmental improvement or deterioration. The revised ES concludes that the impact of the airport on air quality is insignificant in relation to the existing levels of pollution created by road vehicles and is also "insignificant in its own right". It considered that the effect of airborne dust during construction is "likely to have the greatest impact on air quality" and suggested a number of mitigation measures to remedy this effect.

6.3 Valuing Noise and Air Pollution

The following sub-sections examine a selection of monetary evaluation techniques which could, in principle, be used to value the costs of increased noise and air pollution arising from the proposed airport development.

6.3.1 The Contingent Valuation Method

The CVM could be used to estimate the deterioration in environmental quality through increased air and noise pollution. In order for it to provide reliable results, however, the environmental resource under assessment must be of a type which is amenable to its perception by individuals at large. For example, poor air quality can be perceived by individuals through the impairment of visibility, although the level of air pollution would have to be relatively high and comprise pollutants which affect visibility for this to be the case.

The CVM could be applied to evaluate how increased noise levels would affect the local community's enjoyment of an area. The evaluation of noise pollution would have to take into account existing background noise levels, for example, the proximity of the airport to the motorway. The resulting monetary estimate (of increased noise pollution) could be usefully compared to the financial costs of reducing the noise associated with the operation of the airport (eg. mitigation expenditures such as noise insulation measures).

6.3.2 The Travel Cost Method (TCM)

The TCM has been used to estimate air pollution via its impact upon areas used for recreation, and it could be applied to the airport development because of its proposed location upon a site which is, in part, used for recreation. However this could only be conducted in a post-project context, in that it is only feasible to measure the effect of elevated noise and air pollution arising from the airport on local recreational value once the airport is in operation. A significant issue in this context would be whether the increased noise and air pollution arising from the airport's operation is great enough to be perceived by site users to deter them from visiting the area.

6.3.3 The Hedonic Price Method (HPM)

Pearce and Markandya (1989) note that the HPM has been used effectively to evaluate the impact of environmental factors on property values. Studies to date indicate that it may be particularly well suited to estimating the external costs of air and noise pollution on residential areas. A proviso to this success is that the pollution has to be able to be perceived by individuals, and of a quantifiable nature. In addition, the technique has considerable data requirements.

Although the theoretical basis of this technique has been refined over the last decade, the resulting benefit (or cost) estimates still contain a "substantial margin of error in either direction" (Pearce and Markandya, 1989). However, this is not to say that the method is of no value. Pearce and Markandya (1989) recognise that "there are many situations where quantification to within even an order-of-magnitude is a useful tool" in environmental project and policy formulation and analysis. In particular, as long as the weaknesses and strengths of this method are fully understood, such that it is applied only in appropriate situations and carefully appraised, this method can provide very useful indicators of the public's preferences towards specific environmental impacts of developments.

6.3.4 Market Priced Approaches : Noise and Air Pollution

(a) Mitigation Costs

The revised 1991 ES considers a range of noise mitigation actions to reduce the negative effects of airport operation on human beings. These primarily relate to the control of aircraft movements (both in the air and on the ground); the construction of properly insulated buildings for over-night testing and maintenance; the provision of noise insulation "if properties are affected by excessive noise levels in the future" (ES 1991); and the possible establishment of a noise monitoring system "should it be required".

The associated costs of these mitigation actions represent an implicit value for the existing (quieter) environment. This highlights one route through which environmental impacts can be evaluated in monetary terms. The accuracy of these valuations could be checked by comparison with other monetary evaluation techniques. It is important to appreciate the basis of these costs; they are, by design, an underestimate of the full external costs of noise pollution, and their primary use is to enable the magnitude of the costs involved in offsetting an undesirable environmental change to be recognised.

Moreover, if the accuracy of EA information regarding noise levels could be refined (and the increased reliability of such predictions accepted) then the possibility of incurring `superfluous' mitigation costs could be strictly controlled, so increasing the efficiency and cost-effectiveness of both the project and of decision-making. It has been argued that, if it were known through post-project appraisal of operational developments that EA predictions were accurate, implementation of future decisions on similar projects would not need to consider all potential mitigations measures, as only those recommended in the ES would be known to be suitable (Johnson & Bratten 1978, cited in Bisset 1984). However, this is apparently contrary to the SACTRA recommendation to value all mitigation costs, irrespective of whether they are implemented.

The SACTRA report (1992) recommends that "the costs of conservation, mitigation, and clean-up should in any case be explicitly stated in the assessment, even when they are not going to be incurred in fact. This may occur because the agency on whom the cost would fall is not compelled to meet it, and cannot or chooses not to do so. In such a case the agency has decided that the environmental damage is less important than the cost of preventing, or remedying it. In a sense, therefore, an implicit value has been put on the asset in question: the estimated full mitigation cost represents an upper value on the damage in question, the lower bound being zero". The UK Government's official Response to the SACTRA Report (1992) in relation to mitigation costs, documents that "the costs of mitigation, clean-up and protection will always be relevant in decision-making, even if they are not expected to be incurred in fact".
In relation to air pollution, the ES considers that the effect of airborne dust during construction is "likely to have the greatest impact on air quality" and proposed a

selection of mitigation measures. These include, for example, the provision for any contaminated material to be transported in sealed containers and the regular watering of major construction routes within the site, with wheel washing and vehicle spraying to reduce off-site transport dust.

(b) The Dose-Response technique

The dose-response technique is based upon establishing a relationship between a specific amount of pollution (eg. 1 tonne of SO_2/year) and any observed damage it causes (eg. specified amount of damage to crops). This relationship needs to be statistically robust, and this has proved to be the main stumbling block in the successful application of the technique.

Once a dose-response relationship has been established, and statistically tested, the economic costs associated with the unit damages can be calculated.

In relation to air pollution the dose-response approach has been used to calculate the economic costs of the following effects; respiratory illness; death; material damage (e.g buildings); and vegetation damage, such as that caused by aircraft jettisoning surplus fuel. The dose-response technique is less applicable to noise pollution, primarily because documented evidence of its damaging effects (eg. deafness) attributable to specific noise-generating activities is sparse.

6.3.5 Safe Minimum Standard

An underlying principle of this approach is to preserve a diversity of fauna and flora, which represents the safe minimum standard to be preserved into the future. Arguably, the maintenance of air quality is not in itself a fundamental aspect of the technique, (although for example air and water quality will themselves affect the suitability of a given habitat for fauna and flora and so indirectly affect the SMS of individual plants and animals). The SMS approach is therefore not directly applicable for valuing noise or air pollution attributable to the proposed airport development.

7. CONCLUSIONS

The principal objective of this case study is to consider the scope for applying monetary evaluation techniques to the environmental impacts arising from a proposed airport. The work demonstrates that certain techniques could be applied to various environmental assets, as summarised in Table 1.

Table 1 : Application of Monetary Evaluation Techniques

Asset Type	Monetary Evaluation Techniques						
	CVM	TCM	HPM	S-P	D-R	MCs	£ Rev from land
Ancient Woodland	✔	✔	L1	×	_	L	×
Golf Course	L	✔	L1	L	_	✔	✔
Green Belt	✔	✔	L1	L	_	✔	_
Noise Pollution	✔	×	✔	×	L	✔	_
Air Pollution	✔	✔	✔	×	✔	✔	_

Key

-	Not considered in this case study as information not available
✔	Technique applicable
L	Limited applicability in that it is possible to apply this technique for this asset/impact category, but unlikely that it is appropriate in this specific instance. More data is needed before it is possible to decide on the applicability of this technique in this instance.
L1	Unlikely that HPM could isolate importance of the ancient woodland or golf course from other recreational uses of the site.
×	Technique not applicable
CVM	Contingent Valuation Method
TCM	Travel Cost Method
HPM	Hedonic Price Method
S-P	Shadow-Project Method
D-R	Dose-Response Method
MCs	Mitigation Costs
£ Rev from land	= lost financial revenue from loss of this asset (eg. lost golf course revenue due to airport development).

The information in Table 1 relates specifically to the airport proposal, although clearly the ideas are transferable to other situations. The techniques can be roughly divided into those which value the existing environment (i.e without the project), such as the recreational and amenity value of the golf course and other open land, and those which estimate the potential damage associated with impacts arising from the operation of the airport (eg air and noise pollution). Together these values could provide an estimate in monetary terms of the impact of the airport on the natural environment. Set against this conclusion, however, is the fact that valuing the environment in this way is fraught with difficulties. All techniques are hampered by incomplete and/or inadequate data; not all the environmental costs and benefits can be valued and, therefore, included in the financial equation; and some of the values which may be obtained run the risk of being highly misleading. Looked at in this way, it is questionable whether monetary valuation can make a useful contribution to the decision-making process.

The way forward appears to fall somewhere between these two points of view. Monetary evaluation techniques can assist decision-making in the environmental field provided the assumptions made are clearly stated and the person responsible for the decision is aware of all the technical shortcomings.

APPENDIX 6 : SUMMARY OF CASE STUDY FINDINGS

SUMMARY OF CASE STUDY FINDINGS

Results of the Case Studies

1.1 The 11 case studies are a selection of different types of project, planning authority and geographical location, as shown in the Tables 1.1 and 1.2. For confidentiality reasons however, details of the individual case studies are not published. The description of results concentrates on the adequacy of the ES; evaluation of the ES and other environmental information; and the decision-making process. The case studies were taken from the 54 authorities interviewed on the telephone. The aim was to explore the various issues raised in greater detail. It should be noted however that some of the findings in this Appendix repeat those in **Appendix 3**.

Table 1.1 - Breakdown of Case Studies by Authority

Type of Planning Authority	Number of Case Studies
District Councils (England)	3
District Councils (Scotland)	1
District Councils (Wales)	1
Metropolitan Boroughs	1
London Boroughs	1
County Councils (England)	3
National Parks	1
	11

Table 1.2

Type of Planning Authority	Case Study
District (England)	Mixed use development comprising hotel, conference centre, golf course, equestrian facility and timeshare properties.
District (England)	Major sport and leisure complex sponsored by the local authority.
District (England)	Car factory
District (Scotland)	Extension to a quarry.

District (Wales)	Pharmaceutical manufacturing plant including waste incinerator.
Metropolitan Borough	City airport.
London Borough	Redevelopment of a small arms factory for mixed use.
County (England)	Extension to a landfill site.
County (England)	Route improvement to a single carriageway road.
County (England)	Extraction and processing of sand and gravel.
National Park	Extension to a limestone quarry.

1.2 **Status of the Case Study Applications :** One of the criteria for selecting the case studies was that the planning application had been decided. The majority are therefore either refused or approved. However, two do remain undecided.

Table 1.3

Application Status	Number of Case Studies
Approved	7
Refused	2
Pending	2
	11

1.3 Given this particular selection criterion, the majority of the ESs were produced at least two years ago, although they ranged between 1988 and 1993. The more recent ESs are generally those which had accompanied revised planning applications following protracted discussions between the applicants and the authority, or formal requests for additional information. It has been noted in research (e.g. Lee, 1993) that the quality of ESs has improved over time since the implementation of the EC Directive. The case studies therefore do not necessary reflect current practice. At least one of the case study authorities provided evidence to show that the quality of more recent ESs has improved and that their own procedures for dealing with them have benefited from increased experience in handling ESs.

1.4 **Type of Project and Developer :** Table 1.2 describes the type of development proposed in each of the case studies, all of which are 'Schedule 2 projects'. In some cases the ES was volunteered by a developer, whereas in others it was formally requested by the planning authority. In some of the case studies, the local authority promoted the scheme and prepared, or commissioned consultants to prepare, the ES; and in one case the authority issued a development brief to developers who then commissioned the ES. The close involvement of the local authority in a proposed development raises important issues concerning the evaluation of environmental information and the decision-making process which are discussed in more detail in paragraph 1.16 below.

1.5 **Sensitivity of the Site :** All but two of the case studies affected at least one site designated for its conservation value, as listed below.

Table 1.4

Planning/Landscape Designations :	Green Belt Area of High Landscape Value Special Landscape Area Tree Preservation Order	2 1 1 2
Landscape Designations (National)	National Park	1
Conservation Designations :	SSSI (adjacent to site) SSSI (part of site) Ancient Woodland Local Designation	4 1 2 2
Heritage Designations :	Listed Building Ancient Monument	3 1
Other :	High Grade Agricultural Land	2

This reflects the sensitivity of many of the sites, as would be expected for proposals subject to EA.

1.6 **Production of a Scoping Report and Statement of the Methods Used :** All the local authority planning officers interviewed stated that a formal scoping report or method statement was not prepared by the developer for any of the case studies. However, the scope of the EA was discussed between the developer and the planning authority in five cases.

1.7 **Guidance :** Similarly, the local authority planning officers interviewed felt that most developers had made use of Schedule 3 to the EA Regulations as a check-list when preparing an ES. The Schedule was used primarily to determine whether the project was likely to require an EA and to clarify the 'specified' and 'additional' information. Other forms of guidance were also believed to have been used, but in two of the case studies none was mentioned, as shown below:

Table 1.5

Guidance used for preparation of ESs in the Case Studies	Number of Case Studies
The 'Blue Book'	1
The Regulations	6
DoT Manual of Environmental Appraisal	1
County Council Handbook	1
County Surveyors advice on waste disposal	1
None	2

1.8 **Consultation by the Developer :** In seven of the case studies there was limited or no consultation by the developer with local authorities prior to the submission of the ESs. However, with four of the case studies the local authority planning officer stated that there was considerable discussion before and after the application was made. Following the submission of the ES, the developer entered into extensive discussions with statutory consultees in five cases.

1.9 **Assessing the Adequacy of the ES :** None of the ESs considered as part of the case studies was subject to a formal review by the planning officers to assess adequacy. However, one district council did forward an ES to the IEA for formal review, whilst one of the county councils employed a University graduate to prepare an ES evaluation.

1.10 In one of the case studies, an 'Environmental Effects Matrix' was incorporated within the ES which the officers found useful. The likely impacts of each component of the development were identified against eight criteria; surface drainage; agricultural land quality; wildlife habitats; the SSSI; archaeology; public rights of way; tree preservation orders; and public views. An indication of the scale of impact was included; as was an assessment of mitigation opportunities.

1.11 One county council used an established procedure for determining the adequacy of the information contained within an ES. A series of map overlays were produced indicating the main spatial environmental and planning constraints affecting the route, this formed the framework within which the adequacy of the information was judged. In addition the statutory consultees were relied upon to highlight areas of deficiency in the statement.

1.12 Most of the authorities stated that they relied heavily on the statutory consultees to assess the adequacy of the information provided. In one case consultants were commissioned to provide an independent noise assessment to validate the information provided in the ES. However, generally the authorities stated that they did not have sufficient resources to employ consultants, whilst acknowledging that on some of the technical subjects they did not have sufficient expertise in-house. One county council considered that significant advances have been made in their assessment procedures over the last five years both internally and through the standards achieved by developers and their consultants.

1.13 In seven of the case studies the planning authorities stated that they were generally not satisfied with the adequacy of the ES, and requested additional information as a result. In one case a completely new ES was requested. In another case, following the aforementioned review of the ES by the IEA (see para. 1.9 above), the district council requested supplementary information which was also reviewed by the Institute. At this point the ES was considered adequate for its purpose by the authority, based on the IEA recommendation.

1.14 Some of the authorities accepted inadequate ESs, regarding the information presented, and used them only as a starting point for their own internal assessment of the environmental effects. All of the authorities that did this emphasised that, in doing so, they created additional work for themselves, sometimes with quite onerous resource implications. One authority pointed out that the extensive additional work they carried out on the ES not only filled information gaps but was important to satisfy the Committee and the public of the adequacy and the objectivity of the ES.

1.15 The inadequacy of the ES in one case was mainly a result of the developer refusing to supply commercially sensitive information. The authority considered that inadequate information had been submitted to evaluate the ES, and planning permission was subsequently refused.

1.16 Where the local authority prepared the ES or promoted the scheme, the ESs were generally considered adequate by the planning departments. No special procedures were adopted in these cases but there was generally a greater level of consultation during the EA Study and the preparation of the ES. The statutory consultees and public, however, both highlighted what they considered to be inadequacies in these ESs.

1.17 **The Decision-Making Process :** The majority of the planning officers stated that they processed an application accompanied with an ES in the same way as any other development proposal. They were generally sceptical of the use of monetary evaluation techniques to assist this process, citing the following concerns :

- resource implications and technical expertise required;

- limited time available to process an application;

- questionable usefulness of techniques over and above the normal professional judgement of planning officers in weighing up issues;

- difficulty in presenting such techniques to committee members.

1.18 A number of the authorities considered that the ES had been of great importance in the decision-making. In contrast, some of the planning officers interviewed stated that the ES had not contributed greatly to the decision-making process. This partly reflected the inadequacy of the ESs, but was mainly because the negotiations that took place after the planning applications were submitted were considered by the officers to be more important to the decision than the contents of the ESs themselves. Concerns were also expressed that the work needed to ensure that the contents of an ES satisfy EA regulations often duplicates that required for other legislation (eg. site licensing) but without a clear division of responsibility for the eventual decision.

1.19 One district council felt that the ES helped to concentrate the mind of the planning officers on the major environmental issues but it was considered that the assessment of significance of the impacts in the ES was not well thought out and was not wholly in accord with the information presented. The lack of expertise and resources available to the authority, however, meant that the authority had largely to rely on the statutory consultees' comments.

1.20 **Presentation to Planning Committee:** The methods chosen by planning officers to present the information relating to a planning application accompanied by an ES differed between authorities. The particular approach adopted in the case studies depended upon:

- the established methods of presentation for planning applications in general;
- the experience of the authority in dealing with applications accompanied by an ES;
- the sensitivity of the application and the main environmental issues.

1.21 The ES was generally included in the Background Papers available to members, although one authority included the text of the ES as an appendix to the Report to Committee, and another included the non-technical summary from the ES. In three cases, the ES was separately presented to members for consideration.

1.22 The length of the Committee Reports varied considerably. For example, one county council's was limited to 12 pages while one borough council's was over 200 pages. The Committee Reports generally referred to the information included in the ES although officers normally presented their own summaries of the environmental issues involved. One county council included an `Environmental Appraisal' of the scheme prepared by the Principal Planning Officer as an appendix to the Committee Report.

1.23 The Committee Reports included, to a greater or lesser extent, the following information:

- description of the proposals;
- planning history/background;
- planning policy context;
- summaries of consultations, representations and objections;
- officer's discussion/appraisal;
- officer's recommendation.

1.24 In some instances, more elaborate methods of presentation were adopted to complement the Committee Report. For example, one county council displayed drawings of the road scheme under consideration and one district council, when considering the application for a mixed use development (including a hotel conference centre, golf course and timeshare properties), was presented with a video of a similar scheme. In the London borough case study, a large section of the Committee Report was devoted to the amendments that the planning officer considered the developers should make in order to gain approval for the scheme. Copies of representations were not provided in all cases, although in most of the case studies the major grounds of objection were summarised in the Report.

1.25 **Resources :** Generally, the planning authorities found it very difficult to quantify the staff resources used to evaluate the ES as distinct from generally processing the particular application. Three of the authorities stated that some of their staff had had training in EA, but four others raised a concern that they did not consider they had sufficient expertise. In one case, the planning authority employed a consultant to evaluate part of the ES prepared by a consultant, but generally authorities stated that insufficient funds were available. However, four authorities stated they would consider using consultants if technical issues beyond their expertise were raised by a development proposal in the future.

APPENDIX 7 : REVIEW METHODOLOGIES

LEE-COLLEY ENVIRONMENTAL STATEMENT REVIEW PROCEDURE (1990)

The purpose of the Lee-Colley Review Package is to set out a procedure for the review of environmental statements by reviewers who may not possess specialist environmental expertise but who are familiar with the relevant EA regulations and have at least a basic, non-specialist knowledge of EA methodologies and current ideas on good practice in EA. It aims to provide the reviewer with a framework within which to interpret the information contained in an ES; to enable the reviewer to assess the quality and completeness of the information relatively quickly; and to enable the reviewer to make an overall judgement of the acceptability of the ES as a planning document.

The reviewer should thus be able to identify the main strengths and weaknesses of the ES, in particular any omissions which need to be rectified by the provision of further information before impacts can be satisfactorily assessed.

The review is performed using a set of hierarchically arranged Review Topics which are arranged in a pyramidal structure (see **Figure 1**). The reviewer begins at the lowest level, i.e the base of the pyramid, which contains simple criteria relating to specific tasks and procedures. Then, drawing upon these assessments, the reviewer moves progressively upwards from one level to another in the pyramid applying more complex criteria to broader tasks and procedures in the process until the overall assessment of the ES has been completed.

Figure 1

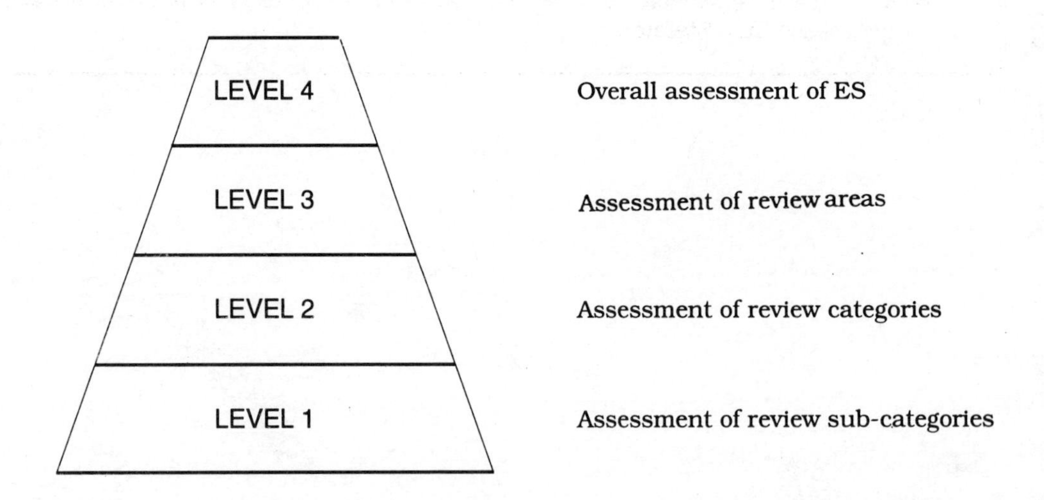

The overall review of the ES (Level 4 in **Figure 1**) is subdivided into four review areas (Level 3) which cover the 4 major areas of EA activity. These are:

1. Description of the development, the local environment, and the baseline conditions.

2. Identification and evaluation of key impacts.

3. Alternatives and mitigation of impacts.

4. Communication of results, i.e. presentation of information in the ES, and the inclusion of a non-technical summary.

Each criterion is graded, as shown in **Table 1**. Letters rather than numbers are used as symbols to discourage reviewers from crude aggregation to obtain assessments at the higher levels in the pyramid.

Symbol	Explanation
A	Relevant tasks well performed, no important tasks left incomplete.
B	Generally satisfactory and complete, only minor omissions and inadequacies.
C	Can be considered just satisfactory despite omissions and / or inadequacies.
D	Parts are well attempted but must, as a whole be considered just unsatisfactory because of omissions and / or inadequacies.
E	Not satisfactory, significant omissions or inadequacies.
F	Very unsatisfactory, important task(s) poorly done or not attempted.
N/A	Not applicable. The Review Topic is not applicable or is irrelevant in the context of this Statement.

CRITERIA TO AID THE REVIEW PROCESS (TOMLINSON 1989)

1.	**Administration**
1.1	Is the ES complete if presented as a series of documents?
1.2	Does the ES address minimum requirements of the EA Regulations? These are: a. a description of the project comprising information on its site, design and size; b. a description of the measures envisaged to avoid, reduce and, if possible, remedy significant adverse effects; c. data required to identify and assess the main environmental effects from the project; d. non-technical summary of the above information.
1.3	Does the ES address issues beyond the minimum requirements of the Regulations?
2.	**Effective Communication**
2.1	Is the purpose and rationale of the project clearly and concisely presented?
2.2	Are the proposed project activities adequately described for readers to become familiar with?
2.3	Is the description of the receiving environment adequate?
2.4	Does the ES present a summary of the significant environmental effects and proposed mitigating measures appropriate for decision-making?
2.5	Is the information presented in a comprehensive format using tables, maps, etc.?
2.6	Are impact predictions easily identified?
2.7	Are all technical terms clearly defined?
3.	**Impact Identification**
3.1	Does the ES discuss those issues considered important by: a. local planning authority; b. statutory consultees; c. expert consultees; d. voluntary organisations; e. local residents; f. general public?
3.2	Are all activities with significant impacts on valued environmental attributes identified?
3.3	Has an environmental assessment method, such as a matrix or checklist, been employed?
3.4	Does the ES address Structure, Local & Subject Plan policies, and Government guidance having a direct bearing on the project, its location and the proposed activities? State why particular environmental attributes and activities are valued.
3.5	Does the ES state why particular environmental attributes and activities are valued?
3.6	Have the timing and duration of specific project activities been described?
3.7	Are the temporal and spatial boundaries for specific environmental effects defined?
3.8	Has the ES identified those individuals, communities and statutory agencies likely to be so affected by the project?
4.	**Alternatives**
4.1	Does the ES present information on the need for the project?
4.2	Have alternative sites, processes, or site layout been addressed?
4.3	Has the "no go" (ie do-nothing) option as a baseline been considered?
5.	**Information Assembly**
5.1	Are sources of information acknowledged, especially expert judgement and opinion?
5.2	Are difficulties including technical deficiencies or lack of know-how encountered in undertaking the assessment recorded?
5.3	Does any basis for questioning the assumptions, data and information used in supporting conclusions exist?
6.	**Baseline Description**
6.1	Does the ES adequately describe: a. existing status of valued environmental resources and land uses? b. site specific planning and environmental policy considerations?
6.2	Does the report identify any variability in: a. natural systems, for example due to seasonal change in habitat use? b. human uses,such as those arising from existing or planned polluting and other land use activities including recreation?
7.	**Impact Prediction**
7.1	Does the ES document the basis on which predictions are made, eg. case studies, models, experts etc?
7.2	Are key variables and assumptions easily identifiable and justified?
7.3	Is there any basis for questioning the validity of predictive techniques employed?
7.4	Is there any undue emphasis on particular issues to the detriment of others, such as construction, operation, abandonment?
7.5	Is each impact considered with a clear and distinct statement of the following: a. variable effected; b. impact magnitude, including its geographical extent; c. timescale of the impact; d. probability; e. significance, including its reversibility; f. confidence to be placed in the prediction.
7.6	Are likely indirect, secondary, cumulative impacts identified?
7.7	Has an attempt been made to isolate project generated impacts from other change resulting from non-project activities and variabilities?
8.	**Mitigation Measures**
8.1	Have the following measures been considered to mitigate adverse environmental effects: a. Could parts of the project be reduced or eliminated? b. Are impacts avoidable or reversible? c. Could affected resources be replaced or compensated for? d. Could environmental management procedures be instituted to reduce adverse effects? e. Could project design, timing, equipment used or site management be modified to reduce effects? f. Could effects be monitored with provision being made for future mitigation as the nature of the impact becomes better known?
8.2	Does the ES present a rationale for selection of chosen mitigating measures?
8.3	Are mitigating measures defined in specific, practical terms and evaluated for their effectiveness and environmental impact?
8.4	Are there any major residual impacts which the ES either justifies why they are acceptable and cannot be mitigated?
9.	**Monitoring and Audits**
9.1	Has a surveillance and monitoring programme been outlined as a specific programme?
9.2	Has the relationship between proposed project monitoring programmes and existing monitoring programmes been outlined?
9.3	Has responsibility for financing, undertaking and reporting the results of monitoring activities been described?
9.4	Does the ES detail a commitment to monitor those impacts considered important,especially where uncertainty exists?
9.5	Does the ES detail a programme of audits to establish a structured learning cycle within the project?

Printed in the United Kingdom by HMSO
Dd 300014 C10 10/94 19585